U0743977

ENTERPRISE STRATEGIC MANAGEMENT
Theory of Sustainable Growth

企业战略管理
持 续 成 长 的 理 论

康健 唐欣 张平 编著

Written and Edited by Kang Jian Tang Xin Zhang Ping

浙江工商大学出版社
ZHEJIANG GONGSHANG UNIVERSITY PRESS

图书在版编目(CIP)数据

企业战略管理:持续成长的理论 / 康健,张平,唐
欣编著. — 杭州:浙江工商大学出版社,2017.6(2025.8重印)
ISBN 978-7-5178-2335-3

Ⅰ. ①企… Ⅱ. ①康… ②张… ③唐… Ⅲ. ①企业管
理－战略管理 Ⅳ. ①F272

中国版本图书馆 CIP 数据核字(2017)第 201338 号

企业战略管理——持续成长的理论

Enterprise Strategic Management—Theory of Sustainable Growth

康健　张平　唐欣　编著

责任编辑	吴岳婷	
封面设计	纪　裕	
责任印制	屈　皓	
出版发行	浙江工商大学出版社	
	(杭州市教工路 198 号　邮政编码 310012)	
	(E-mail:zjgsupress@163.com)	
	(网址:http://www.zjgsupress.com)	
	电话:0571-88904980,88831806(传真)	
排　版	杭州朝曦图文设计有限公司	
印　刷	广东虎彩云印刷有限公司绍兴分公司	
开　本	787mm×1092mm　1/16	
印　张	14.25	
字　数	383 千	
版印次	2017 年 6 月第 1 版　2025 年 8 月第 4 次印刷	
书　号	ISBN 978-7-5178-2335-3	
定　价	55.00 元	

本书受到如下项目资助

(1)湖南省普通高等学校教学改革研究项目:基于 OBE 理念的工商管理专业优势课程群建设研究与实践——以湖南工学院为例(湘教通〔2016〕400 号)

(2)湖南工学院校本级"本科教学工程"项目:战略管理双语教学示范课程(湖工教〔2013〕16 号)

(3)湖南工学院校本级规划教材建设立项项目:战略管理(湖工教〔2015〕49 号)

(4)湖南工学院博士科研启动项目:基于创新链内资源获取的湖南省战略性新兴产业技术创业能力轨迹跃迁研究(HQ16001)

(5)湖南工学院 2016 年大学生研究性学习和创新性实验计划项目:衡阳市能源发展规划

(6)湖南省教育科学"十三五"规划 2017 年度立项课题:双重螺旋,一体两翼:基于项目驱动的地方高校工商管理专业翻转课堂教学模式研究(XJK17BGD011)

2023 年修订说明

本次修订主要增加了财务战略方面的内容。

在第五章增加了基于资本引入的企业分类，包括以经营性资源为主的经营驱动型、以金融性资源为主的债务融资驱动型、以股东入资为主的股东驱动型、以留剩资源为主的利润驱动型以及以均衡利用各类资源为主的并重驱动型。

在第六章增加了财务战略内容，包括企业财务战略分类方法。一种是按职能类型，分为投资战略、筹资战略、营运战略和股利战略；另一种是按综合类型，分为扩展型、稳健型、防御型和收缩型财务战略。还包含财务绩效评估。企业需要及时、准确地披露自己的财务信息，包括财务报表、财务指标等。这些信息可以帮助投资者和利益相关者了解企业的财务状况，提高企业的透明度和信誉度。

在第七章增加了财务与企业战略内容，基于经营性资源的财务分析，企业对于商业信用资源的利用绝不是被动、自然形成的，而是积极主动的，企业往往将最大限度利用商业信用资源作为其优先选择的经营战略与财务战略。基于金融性资源的财务分析，对企业金融性资源的利用或引入状况进行分析和考察可以看出企业集团的财务战略意图和整体战略规划。基于股东留剩资源的财务分析，在一定的盈利规模下，企业可以通过制定不同的股利分配（如现金股利、股票股利或者是二者的组合等）政策，在一定程度上改变企业的财务结构（如改变企业的资产负债率），并对企业的战略特别是融资战略形成支撑。

CONTENTS

Chapter 6　Functional Strategies and Strategic Choices ···················· 99

Chapter 7　Strategy Implementation ·································· 120

Chapter 9   Strategic Evaluation and Control ·························· 179

Chapter 10   Enterprise Growth Strategy ························· 195

Chapter 1   Strategic Management Concept

1.1   Strategic Competitiveness

A strategy of a corporation is a comprehensive plan stating how the corporation will achieve its mission and objectives. 战略是表明一个公司将如何实现使命和目标的综合性计划。

The study of strategic management therefore emphasizes the monitoring and evaluating of external opportunities and threats in light of a corporation's strengths and weaknesses in order to generate and implement a new strategic direction for an organization.

Strategic management is that set of managerial decisions and actions that determines the long-run performance of a corporation. It includes environmental scanning (both external and internal), strategy formulation (strategic planning), strategy implementation, and evaluation and control. 战略管理是决定公司长期绩效的一系列管理决策和行动。包括环境扫描(外部和内部)、战略制定(战略计划)、战略实施,以及评价和控制。

Strategic competitiveness is achieved when a firm successfully formulates and implements a value-creating strategy. A strategy is an integrated and coordinated set of commitments and actions designed to exploit core competencies and gain a competitive advantage. When choosing a strategy, firms make choice among competing alternatives as the pathway for deciding how they will pursue strategic competitiveness. In this sense, the chosen strategy indicates what the firm will do as well as what the firm will not do.

A firm has a competitive advantage when it implements a strategy that creates superior value for customers and that its competitors are unable to duplicate or find too costly to imitate. An organization can be confident that its strategy has resulted in one or more useful competitive advantages only after competitors' efforts to duplicate its strategy have ceased or failed. In addition, firms must understand that no competitive advantage is permanent. The speed with which competitors are able to acquire the skills needed to duplicate the benefits of a firm's value-creating strategy determines how long the competitive advantage will last.

Corporate strategy describe a company's overall direction in terms of its general attitude toward growth and the management of its various businesses and product lines. 企业战略描述了公司的总体方向,对其成长和管理的各种业务和产品线的总体方向。

1

Above-average returns are returns in excess of what an investor expects to earn from other investments with a similar amount of risk. Risk is an investor's uncertainty about the economic gains or losses that will result from a particular investor's uncertainty about the economic gains or losses that will result from a particular investment. The most successful companies learn how to effectively manage risk. Effectively managing risks reduces investors' uncertainty about the results of their investment. Returns are often measured in terms of accounting Figures, such as return on assets, return on equity, or return on sales. Alternatively, returns can be measured on the basis of stock market returns, such as monthly returns (the end-of-the-period stock price minus the beginning stock price, divided by the beginning stock price, yielding a percentage return). In smaller, new venture firms, returns are sometimes measured in terms of the amount and speed of growth rather than more traditional profitability measures because new ventures require time to earn acceptable returns (in the from of return on assets and so forth) on investors' investments.

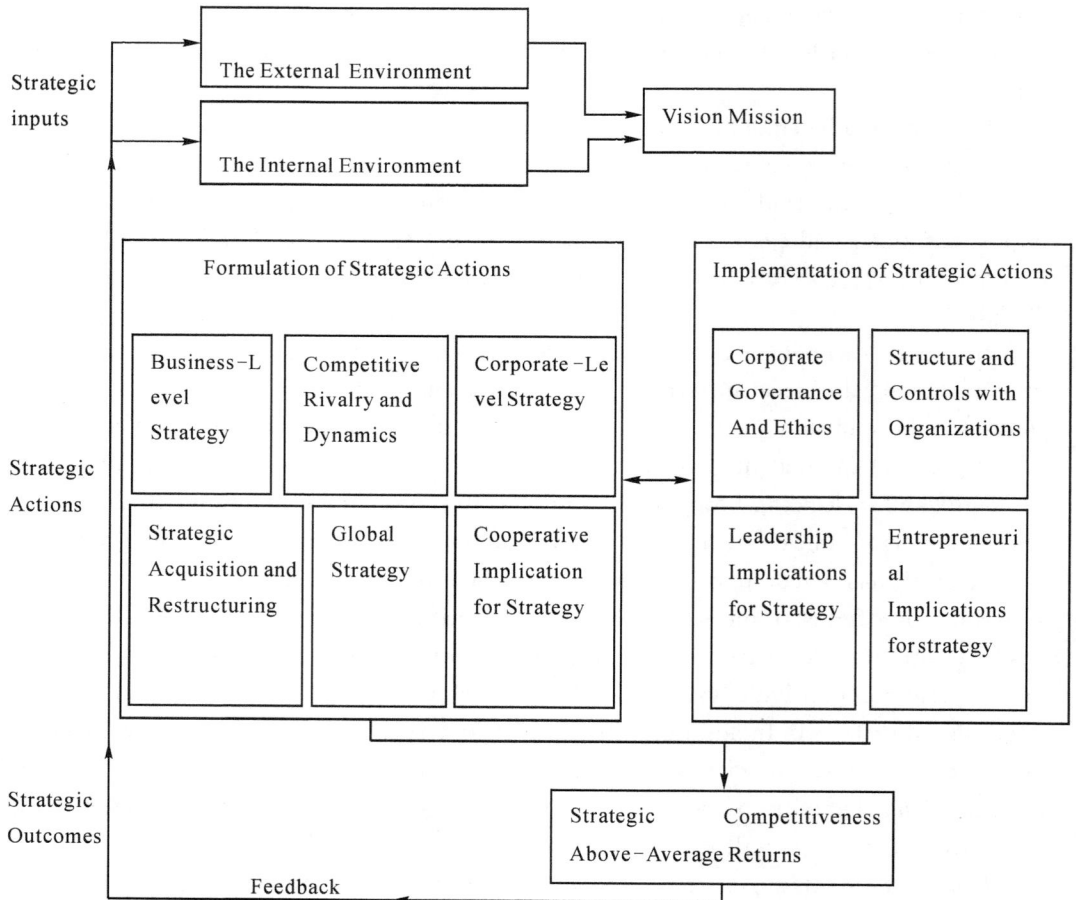

Figure 1. 1

The typical business firm usually considers three types of strategy: corporate, business, and functional. 有代表性的公司经常把战略分为三种类型:公司战略、业务战略和职能战略。

Business strategy usually occurs at the business unit or product level, and it emphasizes improvement of the competitive position of a corporation's products or services in the specific industry or market segment served by that business unit. 业务战略经常发生在业务部门或产品层面,它强调的是一个公司在特定产业或细分市场中通过业务部门提供的服务带来的公司产品或服务中的竞争地位的提升。

Function strategy is the approach taken by a functional area, such as marketing or research and development, to achieve corporate and business unit objectives and strategies by maximizing resource productivity. 职能战略是为了公司和业务单位目标及战略,最大限度地提高资源生产率被企业功能领域,比如市场营销或研开和发展等划分的方法。

A hierarchy of strategy is the grouping of strategy types by level in the organization. 战略的层级是指根据组织的层级对战略类型进行的分组。

1.2　The Global Economy

A global economy is one in which goods, service, people, skills, and ideas move freely move freely across geographic borders. Relatively unfettered by artificial constraints, such as tariffs, the global economy significantly expands and complicates a firm's competitive environment.

1. Study the external environment, especially the industry environment.

The External Environment
- The general environment
- The industry environment
- The competitor environment

2. Locate an industry with high potential for above average returns.

An Attractive Industry
- An industry whose structural Characteristics suggest above-average returns

3. Identify the strategy called for by the attractive industry to earn above average return

Strategy Formulation
- Selection of a strategy linked with above-average returns in a particular industry

4. Develop or acquire assets and skills needed to implement the strategy.

5. Use the firm's strengths(its developed or acquired assets and skills) to implement the strategy.

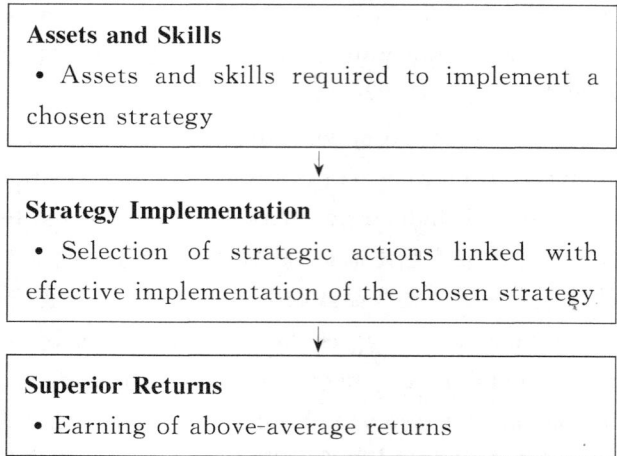

Assets and Skills

• Assets and skills required to implement a chosen strategy

↓

Strategy Implementation

• Selection of strategic actions linked with effective implementation of the chosen strategy

↓

Superior Returns

• Earning of above-average returns

Figure 1. 2

The strategic management model includes a feedback /learning process in which information from each element of the process is used to make possible adjustments to each of the previous elements of the process. 战略管理模型包括一个反馈/学习过程。在这个过程中,来自各个组成部分的信息对前一个过程中的组成部分进行调整。

1.3 Mission

The vision is the foundation for the firm's mission. A mission specifies the business or businesses in which the firm intends to compete and the customers it intends to serve. The firm's mission is more concrete than its vision. However, similar to the vision, a mission should establish a firm's individuality and should be inspiring and relevant to all stakeholders. Together, the vision and mission provide the foundation that the firm needs to choose and implement one or more strategies. The probability of forming an effective mission increases when employees have a strong sense of the ethical standards that guide their behaviors as they work to help the firm reach its vision. Thus, business ethics are a vital part of the firm's discussions to decide what it wants to become (its vision) as well as who it intends to serve and how it desires to serve those individuals and groups (its mission).

Examples: Mission and Vision Statements

Otis Elevator

Our mission: To provide any customer a means of moving people and things up, down, and sideways over short distances with higher reliability than any similar enterprise in the world.

Avis Rent-a-Car

Our business is renting cars. Our mission is total customer satisfaction.

Examples：Mission and Vision Statements（a unique grocery store chain）

Our mission：To give our customers the best food and beverage values that they can find anywhere and to provide them with their formation required for informed buying decisions. We provide these with a dedication to the highest quality of customer satisfaction delivered with a sense of warmth，friendliness，fun，individual pride，and company spirit.

An organization's mission is its purpose or the reason for its existence. It tells what the company is providing to society，such as housecleaning or manufacturing automobiles. A well-conceived mission statement defines the fundamental，unique purpose that sets a company apart from other firms of its type and identifies the scope of the company's operations in terms of products offered and markets served. 使命是组织的目的或存在的理由。一份完美构建的使命陈述诠释了根本而独特的目的，将企业与其他相同类型的公司区别开来，明确了公司产品（包括服务）的经营范围以及所面对的市场。

1.4　Strategic Management Process

The strategic management process is a rational approach firms use to achieve strategic competitiveness and earn above-average returns. Figure 1.1 also features the topics we examine in this book present the strategic management process to you.

This book is divided into three parts. We describe what firms do to analyze their external environment and internal organization. These analyses are completed to identify marketplace opportunities and threats in the external environment and to decide how to use the resources，capabilities，core competencies，and competitive advantages in the firm's internal organization to pursue opportunities and overcome threats. The analyses explained in compose the well-known SWOT analyses（strengths，weaknesses，opportunities，threats）. With knowledge about its external environment and internal organization，the firm forms its strategy taking into account the firm's vision and mission.

The firm's strategic inputs provide the foundation for choosing one or more strategies and deciding how to implement them. As suggested in Figure1.1 by the horizontal arrow linking the two types of implement them. As suggested in Figure 1.1 by the horizontal arrow linking the two types of strategic actions，formulation and implementation must be simultaneously integrated to successfully use the strategic management process. Integration happens as decision makers think about implementation issues when choosing strategies and as they think about possible changes to the firm's strategies while implementing a currently chosen strategy.

We discuss the different strategies firms may choose to use. First, we examine business-level strategies. A business-level strategy describes the actions a firm takes to exploit its competitive advantage over rivals. A company competing in a single product market (e. g. , a locally owned grocery store operating in only one location) has but one business-level strategy while a diversified firm competing in only one location) has but one business-level strategy while a diversified firm competing in multiple product markets (e. g. , General Electric) forms a business-level strategy for each of its businesses. Then we describe the actions and reactions that occur among firms in marketplace competition. Competitors typically respond to and try to anticipate each other's actions. The dynamics of competition affect the strategies firms choose as well as how they try to implement the chosen strategic.

For the diversified firm, corporate-level strategy is concerned with determining the businesses. Other topics vital to strategy formulation, particularly in the diversified company, include acquiring other businesses and, as appropriate, restructuring the firm's portfolio of businesses and selecting an international strategy. With cooperative strategies, firms form a partnership to share their resources and capabilities in order to develop a competitive advantage. Cooperative strategies are becoming increasingly important as firms seek ways to compete in the global economy's array of different markets.

To examine actions taken to implement strategies, we consider several topics in the book. First, we examine the different mechanisms used to govern firms. With demands for improved corporate governance being voiced by many stake-holders in the current business environment, organizations are challenged to learn how to simultaneously satisfy their stakeholders' different interests. Finally, the organizational structure and actions needed to control a firm's operations, the patterns of strategic leadership appropriate for today's firms and competitive environments, and strategic entrepreneurship as a path to continuous innovation are addressed.

It is important to emphasize that primarily because they are related to how a firm interacts with its stakeholders, almost all strategic management process decisions have ethical dimensions. Organizational ethics are revealed by an organization's culture; that is to say, a firm's decisions are a product of the core values that are shared by most or all of a company's managers and employees. Especially in the turbulent and often ambiguous competitive landscape of the twenty-first century, those making decisions as a part of the strategic management process are challenged to recognize that their decisions affect capital market, product market, and organizational stakeholders differently and to regularly evaluate the ethical implications of their decisions. Decision makers failing to recognize these realities accept the risk of placing their firm at a competitive disadvantage with regard to ethical business practices.

As you will discover, the strategic management process examined in this book calls for disciplined approaches to serve as the foundation for developing a competitive advantage. These approaches provide the pathway through which firms will be able to achieve strategic competitiveness and earn above-average returns. Mastery of this strategic management process will effectively serve you, our readers, and the organizations for which you will choose to work.

Strategic management within a firm generally evolves through four sequential phases of development：战略管理过程在公司内部的演变分为四个连续阶段：

Phase 1. Basic financial planning：Seeking better operational control by trying to meet annual budgets. 第 1 阶段　基本财务规划：寻求更好的操作控制以满足年度预算。

Phase 2. Forecast-based planning：Seeking more effective planning for growth by trying to predict the future beyond the next year. 第 2 阶段　以预测为基础的规划：寻求更有效的增长计划，试图预测未来。

Phase 3. Externally oriented strategic planning：Seeking increased responsiveness to markets and competition by trying to think strategically. 第 3 阶段　外部导向型战略规划：寻求增加对市场和竞争的反应，试图认为战略。

Phase 4. Strategic management：Seeking a competitive advantage by considering implementation and evaluation and control when formulating a strategy. 第 4 阶段　战略管理：在制定战略时，考虑实施、评估和控制，寻求竞争优势。

Strategic management consists of four basic elements：(1) environmental scanning, (2) strategy formulation, (3) strategy implementation, and (4) evaluation and control. 战略管理包括四个基本组成部分：(1)环境扫描；(2)战略制定；(3)战略实施；(4)评价和控制。

1.5　Developing a Strategic Vision and Mission

Corrective Adjustments. Why Strategic Management Is a Process. Who Performs the Tasks of Strategy? Benefits of "Thinking and Managing Strategically".

Thinking Strategically. The Three Big Strategic Questions：1. Where are we now? 2. Where do we want to go? (Business to be in and market positions to stake out? Buyer needs and groups to serve? Outcomes to achieve?) 3. How do we get there?

What is Strategy? A company's strategy consists of the set of competitive moves and business approaches that management is employing to run the company Strategy is management's "game plan" to Attract and please customers Stake out a market position Conduct operations Compete successfully Achieve organizational objectives.

What is a Business Model? A company's business model addresses "How do we make money in this business? "Is the strategy that management is pursuing capable of delivering good bottom-line results? Do the revenue-cost-profit economics of the company's strategy make good business sense? Look at the revenue streams the strategy is expected to produce Look at the associated cost structure and potential profit margins Do the resulting earnings streams and ROI indicate the strategy makes sense and that the company has a viable business model?

Strategy vs. Business Model: What is the Difference? Strategy—Deals with a company's competitive initiative sand business approaches Business Model—Concerns whether the revenues and costs flowing from the strategy demonstrate that the business can be amply profitable and viable Strategy Business Model.

Microsoft's Business Model Employ a cadre of highly skilled programmers to develop proprietary code; keep source code hidden from users Employ a cadre of highly skilled programmers to develop proprietary code; keep source code hidden from users Sell resulting operating system and software packages to PC makers and users at relatively attractive prices and achieve large unit sales Sell resulting operating system and software packages to PC makers and users at relatively attractive prices and achieve large unit sales Most costs arise in developing the software; variable costs are small—once breakeven volume is reached, revenues from additional sales are almost pure profit. Most costs arise in developing the software; variable costs are small—once breakeven volume is reached, revenues from additional sales are almost pure profit.

Redhat Linux's Business Model Use volunteer programmers to create the software; make source code open and available to all users Use volunteer programmers to create the software; make source code open and available to all users Give Linux operating system away free of charge to those who download it(charge a small fee to users who want a copy on CD)Give Linux operating system away free of charge to those who download it(charge a small fee to users who want a copy on CD)Make money by employing a cadre of technical support personnel who provide technical support to users for a fee Make money by employing a cadre of technical support personnel who provide technical support to users for a fee.

Why Are Strategies Needed? To proactively shape how a company's business will be conducted To mold their dependent actions and decisions of managers and employees into coordinated, company-wide game plan.

Strategic Management Concept Competent execution of a well-conceived strategy is the best test of managerial excellence and a proven recipe for organizational success! Good Strategy +Good Strategy Execution=Good Management.

Developing a Strategic Vision First Task of Strategic Management Involves thinking strategically about Firm's future business plans Where to "go" Tasks include Creating a roadmap of the future Deciding future business position to stake out Providing long-term direction Giving firm a strong identity

Characteristics of a Strategic Vision A roadmap of a company's future Future technology-product-customer focus Geographic and product markets to pursue Capabilities to be developed Kind of company management is trying to create.

Missions vs. Strategic Visions A strategic vision concerns a firm's future business path—"where we are going" Markets to be pursued Future technology -product-customer focus Kind of company that management is trying to create A mission statement focuses on current business activities—"who we are and what we do" Current product and service offerings Customer needs being served Technological and business capabilities.

Why is a Strategic Vision Important? A managerial imperative exists to look beyond today and think strategically about Impact of new technologies How customer needs and expectations are changing What it will take to outrun competitors Which promising market opportunities ought to be aggressively pursued External and internal factors driving what accompany needs to do to prepare for the future?

A policy is a broad guideline for decision making that links the formulation of strategy with its implementation. Companies use policies to make sure that employees throughout the firm make decisions and take actions that support the corporation's mission, objectives, and strategies. 政策是连接战略制定和实施的决策的广泛纲要。公司利用政策确保全体员工所做出的决策以及采取的行动支持公司的使命、目标和战略。

1.6   Setting Objectives

Types of Objectives Required Financial Objectives Strategic Objectives Outcomes focused on improving financial performance Outcomes focused on improving long-term, competitive business position.

Examples of Financial Objectives Grow earnings per share 15% annually Boost annual return on investment (or EVA) from 15% to 20% within three years Increase annual dividends per share to stockholders by 5% each year Strive for stock price appreciation equal to or above the S&P 500 average Maintain a positive cash flow every year Achieve and maintain a AA bond rating

Examples of Strategic Objectives Increase firm's market share Overtake key rivals on quality or customer service or product performance Attain lower overall costs than rivals Boost firm's reputation with customers Attain stronger foothold in international markets Achieve technological superiority Become leader in new product introductions Capture attractive growth opportunities

Examples: Strategic Objectives Banc One Corporation To be one of the top three banking companies in terms of market share in all significant markets we serve. Domino's

Pizza To safely deliver a hot, quality pizza in 30 minutes or less at a fair price and a reasonable profit.

Objectives are the end results of planned activity. They state what is to be accomplished by when and should be quantified if possible. 短期目标是计划活动的最终结果。短期目标通常明确了各项任务完成的时间和结果，如果可能的话，一般采取量化的形式。

The term goal is often confused with objective. In contrast to an objective，a goal is an open-ended statement of what one wishes to accomplish with no quantification of what is to be achieved and no time frame for completion. 长期目标经常与短期目标混淆。与短期目标相比，长期目标是一种开放式的目标，它没有量化和完成时间的限制。

1.7 Crafting a Strategy

Strategy involves determining whether to Concentrate on a single business or several businesses(diversification)Cater to a broad range of customers or focus on a particular niche Develop a wide or narrow product line Pursue a competitive advantage based on Low cost or Product superiority or Unique organizational capabilities.

Crafting a Strategy Involves deciding how to Respond to changing buyer preferences Respond to new market conditions Grow the business over the long-term The How's That Define a Firm's Strategy. How to grow the business How to please customers How to outcompete rivals. How to respond to changing market conditions. How to manage each functional piece of the business and develop needed organizational capabilities.

Strategic Priorities of McDonald's Continued growth Providing exceptional customer care. Remaining an efficient and quality producer. Developing people at every organizational level. Sharing best practices among all units. Reinventing the fast food concept by fostering innovation in the menu，facilities，marketing，operation and technology.

Core Elements of McDonald's Strategy. Add 1750 restaurants annually Promote frequent customer visits via attractive menu items，low-price specials，and Extra Value Meals. Be highly selective in granting franchises. Locate on sites offering convenience to customers and profitable growth potential. Focus on limited menu and consistent quality. Careful attention to store efficiency. Extensive advertising and use of Mc prefix. Hire courteous personnel pay an equitable wage provide good training.

Crafting Strategy is an Exercise in Entrepreneurship. Strategy-making is a market-driven and customer-driven activity that involves Keen eye for spotting emerging market opportunities. Keen observation of customer needs. Innovation and creativity Prudent risk-taking Strong sense of how to grow and strengthen business

Characteristics of Managers with Good Entrepreneurial Skills Boldly pursue new strategic opportunities. Emphasize out-innovating the competition Lead the way to improve firm performance Willing to be a first-mover and take risks Respond quickly and opportunistically to new developments Devise trail blazing strategies

Why Do Strategies Evolve? There is always an ongoing need to react to Shifting market conditions Fresh moves of competitors. New technologies Evolving customer preferences Political and regulatory changes New windows of opportunity Crisis situations.

Strategic flexibility demands a long-term commitment to the development and nurturing of critical resources. It also demands that the company becomes a learning organization: an organization skilled at creating, acquiring, and transferring knowledge and at modifying its behavior to reflect new knowledge and insights. 战略灵活性需要公司长期努力开发和培育关键性资源,还需要公司成为一个学习型组织——善于创造、获取和转换知识,并不断根据新知识和新见解调整自身行为。

Strategy formulation is the development of long-range plans for the effective management of environmental opportunities and threats, in light of corporate strengths and weaknesses. 战略制定是指基于公司自身的优势和劣势,为有效管理环境中存在的机会和威胁制定企业的长期计划。

1.8   Implementing and Executing a Strategy

What is a Strategic Plan? Where firm is headed—Strategic vision and business mission. Action approaches to achieve targeted results—A comprehensive strategy Short and long term performance targets—Strategic and financial objectives

Strategy Implementation and Execution Strategy implementation and execution is an action-oriented, "make-it-happen" process involving people management, developing competencies and capabilities, budgeting, policy-making, motivating, culture-building and leadership.

What Does Strategy Implementation and Execution Include? Building a capable organization. Allocating resources to strategy-critical activities Establishing strategy-supportive policies Motivating people to pursue the target objectives Tying rewards to achievement of results Creating a strategy-supportive corporate culture Installing needed information, communication, and operating systems. Instituting best practices and programs for continuous improvement. Exerting the leadership necessary to drive the process forward and keep improving.

Strategy implementation is the process by which strategies and policies are put into action through the development of programs, budgets, and procedures. 战略实施是通过

开发各种方案、预算和流程,将战略和政策付诸行动的过程。

Sometimes referred to as operational planning, strategy implementation often involves day-to-day decisions in resource allocation. 战略实施有时称为运营计划,通常涉及资源分配的日常决策。

Monitoring, Evaluating and Taking Corrective Actions as Needed. The tasks of crafting, implementing, and executing a strategy are not a one-time exercise Customer needs and competitive conditions change. New opportunities appear; technology advances; any number of other outside developments occur. One or more aspects of executing the strategy may not be going well New managers with different ideas take over Organizational learning occurs All these trigger the need for corrective actions and adjustments Fifth Task of Strategic Management

Characteristics of the Strategic Management Process. Need to do the five tasks never goes away Boundaries among the five tasks are blurry Strategizing is not isolated from other managerial activities Time required comes in lumps and spurts. The big challenge : To get the best strategy-supportive performance from employees, perfect current strategy, and improve strategy execution.

A program is a statement of the activities or steps needed to accomplish a single-use plan. It makes the strategy action oriented. 方案是完成某个单一用途计划所必需的各种活动或步骤的详细说明,有助于使战略以行动为导向。

Research has revealed that organizations that engage in strategic management generally outperform those that do not. The attainment of an appropriate match or "fit" between an organization's environment and its strategy, structure, and processes has positive effects on the organization's performance. 研究表明,进行战略管理的组织,其绩效通常优于那些没有进行战略管理的组织。组织的战略、结构和流程与环境之间的匹配或"适合",对于实现预期绩效有积极的影响。

1.9　Approaches to Performing the Strategy-Making Task

Chief Architect Manager personally functions as chief strategist Delegate-It -to-Down-the-Line Managers Manager delegates some strategy-making responsibility to subordinates in charge of key organizational units. Collaborative Team Manager enlists assistance and advice of key subordinates in hammering out a consensus strategy Corporate Entrepreneur Manager encourages subordinates to develop and champion proposals for new ventures

Strategic Role of a Board of Directors Critically appraise and ultimately approve strategic action plans. Evaluate strategic leadership skills of the CEO and candidates to succeed the CEO.

Strategic Management Principle. A board of director's role in the strategic management process is to critically appraise and ultimately approve strategic action plans and to evaluate the strategic leadership skills of the CEO and others in line to succeed the incumbent CEO.

Benefits of "Strategic Thinking" and a " Strategic Approach" to Managing Guides entire firm regarding what it is we are trying to do and to achieve" Makes managers more alert to winds of change, new opportunities, and threatening developments Unifies numerous strategy-related decisions and organizational efforts. Creates a proactive atmosphere. Promotes development of an evolving business model focused on bottom-line success Provides basis for evaluating competing budget requests.

Environmental scanning is the monitoring, evaluating, and disseminating of information from the external and internal environments to key people within the corporation. 环境扫描是指监测和评估公司内外部环境,并向内部重要成员传递信息的过程。

A budget is a statement of a corporation's programs in dollar terms. Used in planning and control, it lists the detailed cost of each program. 预算是以货币计量的财务报表形式阐述公司的方案。预算应用于计划和控制领域,罗列出每个项目的详细成本。

According to Henry Mintzberg, the most typical strategic decision making modes are entrepreneurial, adaptive, and planning. 根据亨利·明茨伯格的研究,最典型的战略决策模式包括企业家模式、适应性模式和规划模式。

Entrepreneurial mode: In this mode of strategic decision making, the strategy is developed by one powerful individual. 企业家模式:在这种战略决策模式中,战略是被一种强有力的个体所推动的。

Adaptive mode: Sometimes referred to as "muddling through," this decision making mode is characterized by reactive solutions to existing problems, rather than a proactive search for new opportunities. 适应性模式:有时被称为"得过且过",这种决策模式以反应存在的问题为解决方案为特征,而不是对新的机会做积极研究。

Planning mode: This decision-making mode involves the systematic gathering of appropriate information for situation analysis, the generation of feasible alternative strategies, and the rational selection of the most appropriate strategy. 规划模式:这一决策模式包括对形势分析的系统收集、可行的替代战略的产生以及对最合适战略的合理选择。

Good arguments can be made for using either the entrepreneurial or adaptive modes or planning mode in certain situations. 在某些特定情况下,无论使用企业家模式还是适应性模式或规划模式,都有很好的理由。

The planning mode includes the basic elements of the strategic management process, is a more rational and thus better way of making strategic decisions. 规划模式包括战略管理过程的基本组成部分,是更加理性,也是更好的战略决策方式。

1.10　Strategic Management Principle

Effective strategy-making begins with a vision of where the organization needs to head! Example: John Deere's Strategic Vision Who Are We? John Deere has grown and prospered through a long-standing partnership with the world's most productive farmers. Today, John Deere is a global company with several equipment operations and complementary service businesses. These businesses are closely interrelated, providing the company with significant growth opportunities and other synergistic benefits.

Procedures, sometimes termed standard operating procedures, are a system of sequential steps or techniques that describe in detail how a particular task or job is to be done. They typically detail the various activities that must be carried out for completion of a corporation's program. 流程有时称为标准作业流程，是一个连续的步骤或技术的系统，详细描述了如何完成一项特定任务或工作。流程通常必须详细描述完成公司方案所必需的各种活动。

Evaluation and control is the process by which corporate activities and performance results are monitored so that actual performance can be compared with desired performance. 评价和控制是指监控公司各种活动和绩效的结果，从而保证实际绩效与期望绩效之间具有可比性的过程。

Example：John Deere's Strategic Vision Where Are We Going? Deere is committed to providing genuine evaluate to the company's stakeholders. In support of that commitment, Deere aspires to: Grow and pursue leadership positions in each of our businesses. Extend our preeminent leadership position in the agricultural equipment market worldwide. Create new opportunities to leverage the John Deere brand globally.

Example：John Deere's Strategic Vision How Will We Get There? By pursuing the broader corporate goals of profitable growth and continuous improvement, each of the company's businesses is expected to: Achieve world-class performance by attaining a strong competitive position in target markets. Exceed customer expectations for quality and value. Earn in excess of the cost of capital over a business cycle.

Example：John Deere's Strategic Vision How Will We Get There? By growing profitably and continuously improving, each of the company's businesses will benefit from and contribute to Deere's unique intangible assets: Our distinguished brand. Our heritage of integrity and team work. Our advanced skills. The special relationships that have long existed between the company and our employees, customers, dealers and other business partners around the world.

Example：John Deere's Strategic Vision. How Will We Measure Our Performance? Each business will make a positive contribution to the corporation's objectives in the pursuit of creating genuine value for our stakeholders. Our "scorecard" includes: Human Resources—employee satisfaction, training Customer Focus—loyalty, market leadership Business Processes—productivity, quality, cost, environment Business Results—return on assets, sales growth.

Characteristics of a Mission Statement. Defines current business activities Highlights boundaries of current business Conveys Who we are, What we do, and Where we are now Company specific, not generic—so as to give a company its own identity. A company's mission is not to make a profit! The real mission is always—"What will we do to make a profit?"

Defining a Company's Business. A good business definition incorporates three factors Customer needs—What is being satisfied Customer groups—Who is being satisfied Technologies and competencies employed—How value is delivered to customers to satisfy their needs.

The distinguishing characteristic of strategic management is its emphasis on strategic decision making. 战略管理的显著特征是强调战略决策。

Unlike many other decisions, strategic decision deal with the long-run future of the entire organization and have three characteristics: Rare, Consequential, Directive. 不同于其他决策,战略决策面对的是整个组织的未来,主要有三个特征:稀缺性、重要性、指导性。

Business Mission: Cardinal Health Cardinal Health is a leading provider of services supporting health care worldwide. The company offers a broad array of services for health-care providers and manufacturers to help them improve the efficiency and quality of health care. These services include pharmaceutical distribution, health-care product manufacturing and distribution, drug delivery systems development... retail pharmacy franchising, and health-care information systems development.

Business Mission: JDS Uniphase is the leading provider of advanced fiber optic components and modules. These products are sold to the world's leading telecommunications and cable television system providers... Our products perform both optical-only functions and optoelectronic functions within fiber option networks. Our products include semiconductor lasers, and isolators for fiber optic applications. In addition, we design, manufacture, and market laser subsystems for a broad range of OEM applications.

Business Mission: Russell Corp. Russell Corporation is a vertically integrated international designer, manufacturer, and marketer of athletic uniforms,..., and a comprehensive line of light weight, yarn-dyed woven fabrics. The Company's manufacturing operations include the entire process of converting raw fibers into finished apparel and fabrics. Products are marketed to sporting goods dealers, department and specialty stores, mass merchandisers, ..., and other apparel manufacturers.

Broad or Narrow Mission Statements? Narrow enough to specify real arena of interest Serve as Boundary for what to do and not do Beacon of where top management intends to take firm Diversified companies have broader business definitions than single-business enterprises.

Definitions: Broad vs. Narrow Scope Broad Definition Furniture Telecommunications Beverages Global mail delivery Travel & tourism Narrow Definition Wrought-iron lawn furniture. Long-distance telephone service. Soft drinks Overnight package delivery Caribbean cruises

Business Mission: The McGraw Hill Companies (a diversified firm) The McGraw-Hill Companies is a global publishing, financial, information and media services company with such renowned brands as Standard & Poor's, Business Week, and McGraw-Hill educational and professional materials. The Company provides information via various media platforms: books, magazines and newsletters; on-line; via television, satellite and FM sideband broadcast; and software, videotape, facsimile and CD-ROM products. The Company now creates more than 90% of its information on digital platforms and its business units are represented on more than 75 Web sites.

Business Mission: FDX Corporation(a diversified firm)FDX is composed of a powerful family of companies: FedEx, RPS, Viking Freight, FDX Global Logistics and Roberts Express. These companies offer logistics and distribution solutions on a regional, national and global scale: fast, reliable, time-definite express delivery; ... expedited same-day delivery; ...; and integrated information and logistics solutions With all this expertise under one umbrella, the FDX companies can provide businesses with the competitive advantage they need by providing streamlined solutions that are on the cutting edge of technology.

Example: Mission Statement The Gillette Company Our mission is to achieve or enhance clear leadership, worldwide, in the existing or new core consumer product categories in which we choose to compete. Current core categories are: Male grooming products-blades and razors, electric shavers, shaving preparations and deodorants... Female grooming products-wet shaving products, hair removal and hair care appliances and deodorants... Alkaline and specialty batteries and cells. Writing instruments and correction products. Certain areas of the oral care market- toothbrushes... Selected areas of the high-quality small household appliance business-coffeemakers.

Mission Statements for Functional Departments Spotlights department's Role and scope of activities Direction which department needs to pursue Contribution to firm's overall mission.

Mission Statements of Functional Departments HUMAN RESOURCES To contribute to organizational success by developing effective leaders, high performance teams, and maximizing the potential of individuals. CORPORATE SECURITY to provide services for the protection of corporate personnel and assets through preventive measures and investigations.

Questions to Address in Developing a Strategic Vision1. What changes are occurring in the market arena(s) where we operate and what implications do these changes have for our future direction? 2. What new or different customer needs should we be moving to satisfy? 3. What new or different buyer segments should we be concentrating on? 4. What new geographic or product markets should we be pursuing? 5. What should the company's business makeup look like in 5 years? 6. What kind of company should we be trying to become?

Entrepreneurial Challenges in Forming a Strategic Vision How to creatively prepare a company for the future How to keep the company responsive to Evolving customer needs Competitive pressures. New technologies New market opportunities Growing or shrinking opportunities.

Intel's "Strategic Inflection points" prior to mid-1980s focused on memory chips. It abandoned memory chip business and became a preeminent supplier of microprocessors to PC industry. Thus, it made PC a central appliance in workplace and home and became the undisputed leader in driving PC technology forward. In 1998, it shifted focus from PC technology to becoming the preeminent building block supplier to the Internet economy.

Communicating the Vision. An exciting, inspirational vision Challenges and motivates workforce Arouses strong sense of organizational purpose Induces employee buy-in Galvanizes people to live the business

Managerial Value of a Well-Conceived Strategic Vision and Mission Crystallizes long-term direction. Reduces risk of rudderless decision-making Conveys organizational purpose and identity Keeps direction-related actions of lower-level managers on common path Helps organization prepare for the future.

Establishing Objectives Second Direction-Setting Task Represent commitment to achieve specific performance targets by a certain time Should be stated in quantifiable terms and contain a deadline for achievement Spell-out how much of what kind of performance by when.

Purpose of Objective-Setting Substitutes results-oriented decision-making for aimlessness over what to accomplish. Provides a set of benchmarks for judging organizational performance.

Two Types of Objectives Are Required Financial Objectives Strategic Objectives Outcomes that improve firm's financial performance. Outcomes that strengthen a firm's competitiveness and long-term market position.

Example: Corporate Objectives To make all our companies leaders in their industries in quality while exceeding customer expectations. To achieve a 50% share of the U. S. beer market. To establish and maintain a dominant leadership position in the international beer market. To provide all our employees with challenging and rewarding work, ..., and opportunities for personal development, advancement, and competitive compensation. To provide our shareholders with superior returns by achieving double-digit annual earnings per share growth, Anheuser-Busch(strategic & financial objectives).

Example: Corporate Objectives Exodus Communications (strategic objectives) Extend our market leadership and position Exodus as the leading brand name in the category. Enhance our systems and network management and Internet technology services. Accelerate our domestic and international growth. Leverage our technical expertise to address new market opportunities in e-commerce.

Strategic or Financial Objectives—Which Take Precedence? Pressures for better short-term financial performance become pronounced when Firm is struggling financially Resource commitments for new strategic initiatives may hurt bottom-line for several years. Proposed strategic moves are risky. Otherwise strategic objectives merit top priority—a firm that consistently passes up opportunities to strengthen its long-term competitive position. Risks diluting its competitiveness Risks losing momentum in its markets. Hurts its ability to fend off rivals' challenges.

Strategic Management Principle. Building a stronger long-term competitive position benefits shareholders more lastingly than improving short-term profitability!

Concept of Strategic Intent. A company exhibits strategic intent when it relentlessly pursues an ambitious strategic objective and concentrates its competitive actions and energies on achieving that objective!

Characteristics of Strategic Intent Indicates firm's intent to stake out a particular position over the long-term. Involves establishing a BHAG—"big, hairy, audacious goal" Signals relentless commitment to winning.

Short-Range Versus Long-Range Objectives Short-Range objectives. Targets to be achieved soon. Serve as stair steps for reaching long-range performance. Long-Range objectives Targets to be achieved within 3 to 5 years Prompt actions now that wil permit reaching targeted long-range performance later.

Objectives Are Needed at All Levels Objective-setting process is top-down, not bottom-up! 1. First, establish organization-wide objectives and performance targets. Next, set business and product line objectives. Then, establish functional and departmental objectives4. Individual objectives are established last.

Strategic Management Principle Objective-setting needs to be more of atop-down than a bottom-up process in order to guide lower-level managers and organizational units toward outcomes that support the achievement of overall business and company objectives.

Crafting a Strategy Third Direction-Setting Task An organization's strategy deals with How to make the strategic vision a reality and achieve target objectives. The game plan for Pleasing customers Conducting operations Building a sustainable competitive advantage. Strategy constitutes management's business model for producing good profitability.

Strategizing Involves HOW to... Achieve performance targets Out-compete rivals and achieve a sustainable competitive advantage Respond to changing market conditions and new customer requirements. Make the strategic vision a reality Our game plan for running the company will be. Characteristics of Strategy-Making Strategy is action-oriented Strategy evolves over time Strategy-making is an ever-ending, ongoing task.

Rule-Breaking Strategies Challenge fundamental conventions by Reconceiving a product or service (Creating a single-use disposable camera) Redefining the marketplace (Detouring retailers by selling online at the company's website) Redrawing industry boundaries (Getting credit cards from Shell Oil or General Motors or AOL).

Tasks of Corporate Strategy Moves to achieve diversification. Actions to boost performance of individual businesses Capturing valuable cross-business strategic fits that result in $1+1=3$ effects! Establishing investment priorities and steering corporate resources into the most attractive businesses

What Business Strategy Involves Forming responses to changes in industry and competitive conditions, buyer needs and preferences, economy, regulations, etc. Crafting competitive moves to produce sustainable competitive advantage. Building competitively valuable competencies and capabilities. Uniting strategic initiatives of functional areas Addressing strategic issues facing the company.

Functional Strategies Game plan for a strategically-relevant function activity, or business process. Details how key activities will be managed Provide support for business strategy Specify how functional objectives are to be achieved.

Operating Strategies Concern narrower strategies for managing grassroots activities and strategically-relevant operating units. Add detail to business and functional strategies.

Example: Operating Strategy Improving Delivery & Order-Filling Manufacturer of plumbing equipment emphasizes quick delivery and accurate order-filling ask its customer service approach. Warehouse manager took following approaches: Inventory stocking strategy allowing 99% of all orders to be completely filled without backordering any item Staffing strategy of maintaining workforce capability to ship any order within 24 hours.

Example: Operating Strategy Boosting Worker Productivity. To boost productivity by 10%, managers of firm with low-price, high-volume strategy take following actions: Recruitment manager develops selection process designed to weed out all but best-qualified candidates. Information systems manager devises way to use technology to boost

19

productivity of office workers. Compensation manager devises improved incentive compensation plan Purchasing manager obtains new efficiency -increasing tools and equipment.

Uniting the Company's Strategy-Making Effort. A company's strategy is a collection of strategies and initiatives being acted on by managers at various organizational levels. Separate levels of strategy must be unified into a cohesive, company-wide action plan Pieces of strategy should fit together like the pieces of a puzzle.

What Do We Mean by "Corporate Social Responsibility?" Conducting company activities within bounds of what is considered ethical and in public interest. Responding positively to emerging societal priorities and expectations. Demonstrating willingness to take need education ahead of regulatory confrontation. Balancing stockholder interests against larger interest of society as a whole Being a "good citizen" in community.

Competitive Conditions and Industry Attractiveness. A company's strategy has to be responsive to Fresh moves of rival competitors Changes in industry' spruce-cost-profit economics. Shifting buyer needs and expectations. New technological developments Pace of market growth.

Strategic Management Principle. A company's strategy can't produce real market success unless it is well-matched to industry and competitive conditions! Company Opportunities and Threats For strategy to be successful, it has to Be well matched to capturing a company's best opportunities. And help counteract threats to the company's well-being.

Company Strengths, Competencies, and Competitive Capabilities A company must have or be able to acquire the resources, competencies, and competitive capabilities needed to execute the chosen strategy. Resource deficiencies, gaps in skills, and weaknesses in competitive position make pursuit of certain strategies risky or altogether unwise.

Strategic Management Principle A company's strategy ought to be grounded in its resource strengths and in what it is good at doing (its competencies and competitive capabilities); it is perilous to discount the competitive liabilities of company's resource deficiencies and skills gaps.

Ambitions, Philosophies, and Ethics of Key Executives Managers generally stamp strategies they craft with their own personal. Ambitions Values Business philosophies Attitudes toward risk Ethical belief.

Shared Values and Company Culture Values and culture often shape the strategic moves a company will Consider Reject. It is generally unwise for a company to undertake strategic moves which conflict with Its culture Values widely shared by managers and employees.

Hewlett-Packard's Basic Values: "The HP Way" Sharing firm's success with employees Showing trust and respect for employees. Providing customers with products or services of the greatest value. Being genuinely interested in providing customers with effective solutions to their problems. Making profit a high stockholder priority. Avoiding use of long-term debt to finance growth Individual initiative, creativity, &.teamwork Being a good corporate citizen

Linking Strategy With Ethics Ethical and moral standards go beyond Prohibitions of law and Language of "thou shaft not" Ethical and moral standards involve Issues of duty and Language of "should and should not do".

A Firm's Ethical Responsibilities to Its Stakeholders Owners/ shareholders – Rightfully expect some form of return on their investment. Owners/ shareholders – Rightfully expect some form of return on their investment Employees-Rightfully expect respect for their worth and devoting their energies to firm Employees-Rightfully expect respect for their worth and devoting their energies to firm Customers-Rightfully expect a seller to provide them with a reliable, safe product or service. Customers-Rightfully expect a seller to provide them with a reliable, safe product or service. Suppliers-Rightfully expect to have an equitable relationship with firms they supply. Community-Rightfully expect businesses to be good citizens in their community.

Chapter 2   External Environment Analysis

Things are always different—the art is figuring out which differences matter.

What Is Situation Analysis? Two considerations Company's external or macro-environment Industry and competitive conditions Company's internal or micro-environment Competencies, capabilities, resource strengths and weaknesses, and competitiveness

Environmental scanning is the monitoring, evaluating, and disseminating of information from the external and internal environments to key people within the corporation. 环境扫描是监测、评估内外部环境,并将这些信息传达给公司内部关键人物的过程。

Strategic Thinking and Analysis Leads to Good Strategic Choices1. Industry's dominant economic traits. Nature of competition & strength of competitive forces. Drivers of industry change. Competitive position of rivals. Strategic moves of rivals. Key success factors. Conclusions about industry attractiveness. Assess Industry & Competitive Conditions. Assessment of company's present strategy. Resource strengths and weaknesses, market opportunities, and external threats. Company's costs compared to rivals. Strength of company's competitive position. Strategic issues that need to be addressed. Assess Company Situation Identify Strategic Options for the Company Select the Best Strategy for the Company.

A corporation's scanning of the environment should include analyses of all the relevant elements in the task environment. 扫描一个公司的环境应该包括分析任务环境中的所有相关要素。

This is known as strategic myopia: the willingness to reject unfamiliar as well as negative information. 战略性近视是指乐于拒绝陌生以及负面的信息。

Key Considerations Regarding the Industry and Competitive Environment Industry's dominate economic traits Competitive forces and strength of each force Drivers of change in their industry Competitor analysis Key success factors Conclusions: Industry attractiveness.

What are the Industry's Dominant Economic Traits? Market size and growth rate Scope of competitive rivalry Number of competitors and their relative sizes Prevalence of backward/forward integration Entry/exit barriers Nature and pace of technological change. Product and customer characteristics Scale economies and experience curve effects. Capacity utilization and resource requirements Industry profitability.

The Experience Curve Effect An experience curve exists when a company's unit costs decline as its cumulative production volume increases because of accumulating production know-how. Growing mastery of the technology. The bigger the experience curve effect, the bigger the cost advantage of the firm with the largest cumulative production volume

Relevance of Key Economic Features Economic Feature Market Size Market growth rate Capacity surpluses/shortages Industry profitability Entry/exit barriers. Product is big-ticket item for buyers. Standard products Rapid technological change. Capital requirements Vertical integration Economies of scale Rapid product innovation. Strategic Importance Small markets don't tend to attract new firms; large markets attract firms looking to acquire rivals with established positions in attractive industries. Fast growth breeds new entry; slow growth spawns increased rivalry& shake-out of weak rivals. Surpluses push prices& profit margins down; shortages pull them up High-profit industries attract new entrants; depressed conditions lead to exit. High barriers protect positions and profits of existing firms; low barriers make existing firms vulnerable to entry. More buyers will shop for lowest price Buyers have more power because it's easier to switch from seller to seller Raises risk; investments in technology facilities/equipment may become obsolete before they wear out. Big requirements make investment decisions critical; timing becomes important; creates a barrier to entry and exit. Raises capital requirements; often creates competitive &cost differences among fully vs. partially vs. non-integrated firms. Increases volume& market share needed to be cost competitive Shortens product life cycle; increases risk because of opportunities for leapfrogging.

Analyzing the Five Competitive Forces: How to Do It Assess strength of each of the five competitive forces (Strong? Moderate? Weak?) Rivalry among competitors Competition from substitute products Competitive threat from potential entrants. Bargaining power of suppliers and supplier-seller collaboration. Bargaining power of buyers and buyer-seller collaboration. Explain how each force acts to create competitive pressure—What are the factors that cause each force to be strong or weak? Decide whether overall competition(the combined effect of all five competitive forces)is brutal, fierce, strong, normal/moderate, or weak.

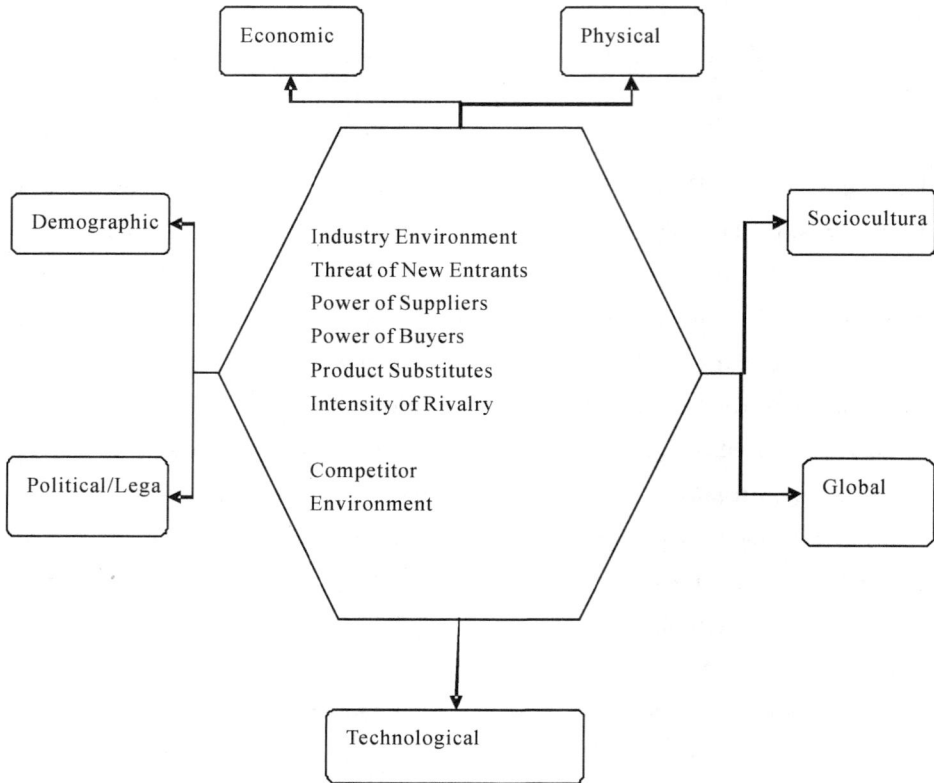

Figure 2. 1　The External Environment

2. 1　The General，Industry，and Competitor Environments

The general environment is composed of dimensions in the broader society that influence an industry and the firms within it. We group these dimensions into seven environmental segment：demographic，economic，political/legal，socio-cultural，technological，global，and physical. Examples of elements analyzed in each of these segments are shown in Table 2. 1.

Table 2. 1　The General Environment：Segments and Elements

Demographic segment	• Population size • Age structure • Geographic distribution	• Ethnic mix • Income distribution

Economic segment	• Inflation rates • Interest rates • Trade deficits or surpluses • Budget deficits or surpluses	• Personal savings rate • Business savings rates • Gross domestic product
Political/Legal segment	• Antitrust laws • Taxation laws • Deregulation philosophies	• Labor training laws • Educational philosophies and policies
Sociocultural segment	• Women in the workforce • Workforce diversity • Attitude about the quality of work life	• Shift in work and career preferences • Shift in preferences regarding product and service characteristics
Technological segment	• Product innovations • Applications of knowledge	• Focus of private and government-supported R&D expenditures • New communication technologies
Global segment	• Important political events • Critical global markets	• Newly industrialized countries • Different cultural and institutional attributes
Physical environment segment	• Energy consumption • Practices used to develop energy sources • Renewable energy efforts • Minimizing a firm's environmental footprint	• Availability of water as a resource • Producing environmentally friendly products • Reacting to natural or man-made disasters

Industry environment analysis refers to an in-depth examination of key factors within a corporation's task environment. 行业环境分析是指全面深入地检验公司任务环境中的关键驱动力量。

The natural environment，the societal environment，the task environment should be analyzed in external environment. 自然环境、社会环境、任务环境应该在企业外部环境中被分析。

The natural environment includes physical resources，wildlife，and climate that are an inherent part of existence on Earth. These factors form an ecological system of interrelated life. 自然环境包括物理资源、野生生物和气候，是地球上的固有存在。这些因素构成具有内在关联的生态系统。

The societal environment is mankind's social system that includes general forces that do not directly touch on the short-run activities of the organization that can, and often do, influence its long-run decisions. These forces involve economic forces, technological forces, political-legal forces, sociocultural forces. 社会环境是人类的社会制度，包括各种

一般驱动力量。这些驱动力量不会直接触及组织的短期活动,但往往会影响组织的长期决策。这些驱动力量包括经济驱动力量、技术驱动力量、政治(法律)驱动力量、社会文化驱动力量。

The task environment includes those elements or groups that directly affect the corporation and, in turn, are affected by it. 任务环境包括那些直接影响公司,反过来又被公司影响的驱动力量或群体。

Firms cannot directly control the general environment's segments. The recent bankruptcy filings by General Motors and Chrysler Corporation highlight this fact. These firms could not directly control various parts of their external environment, including the economic and political/legal segments; however, these segments are influencing the actions the firms are taking, including Chrysler's alliance with Fiat. Because firms cannot directly control the segments of their external environment, successful ones learn how to gather the information needed to understand all segments and their implications for selecting and implementing the firm's strategies.

The industry environment is the set of factors that directly influences a firm and its competitive actions and responses: the threat of new entrants, the power of suppliers, the power of buyers, the threat of product substitutes, and the intensity of rivalry among competitors. In total, the interactions among these five factors determine an industry's profit potential; in turn, the industry's profit potential influences the choices each firm makes about its strategic actions. The challenge for a firm is to locate a position within an industry where it can favorably influence the five factors or where it can successfully defend against their influence. The greater a firm's capacity to favorably influence its industry environment, the greater the likelihood that the firm will earn above-average returns.

How companies gather and interpret information about their competitors is called competitor analysis. Understanding the firm's competitor environment complements the insights provided by studying the general and industry environments. This means, for example, that BP wants to learn as much as it can about its major competitors-such as Exxon-Mobil and Royal Dutch Shell plc-while also learning about its general and industry environments.

Analysis of the general environment is focused on environmental trends while an analysis of the industry environment is focused on the factors and conditions influencing an industry's profitability potential and an analysis of competitors is focused on predicting competitors' actions, responses, and intentions. In combination, the results of these three analyses influence the firm's vision, mission, and strategic actions. Although we discuss each analyses of the general environment, the industry environment, and the competitor environment.

An industry is a group of firms producing a similar product or service, such as financial services or soft drinks. 行业是指生产类似产品或服务的公司集群,如金融服务或软饮料。

Michael Porter, an authority on competitive strategy, contends that a corporation is most concerned with the intensity of competition within its Industry. 竞争战略的学术权威迈克尔·波特认为,公司最关心的是所处行业内的竞争强度。

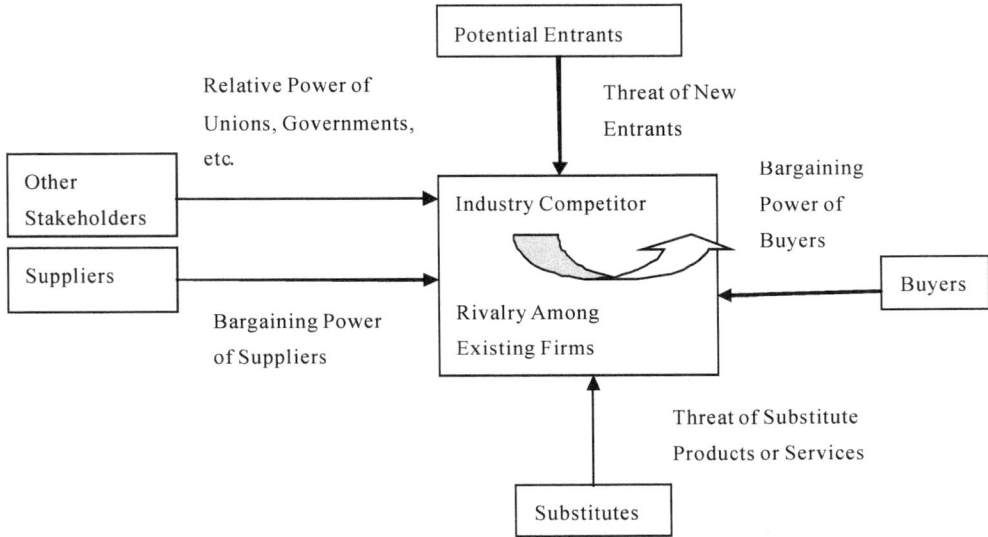

Figure 2.2

In carefully scanning its industry, the corporation must assess the importance to its success of each of the following six forces: threat of new entrants, rivalry among existing firms, threat of substitute products, bargaining power of buyers, bargaining power of suppliers, and relative power of other stakeholders. 在认真扫描行业的过程中,公司必须评估以下六个因素对于成功的重要性:新进入者的威胁、现有企业之间的竞争对抗性、替代产品的威胁、购买者的议价能力、供应商的议价能力、其他利益相关者的相对力量。

2.2　External Environmental Analysis

Most firms face external environments that are highly turbulent, complex, and global-conditions that make interpreting those environments difficult. To cope with often ambiguous and incomplete environmental data and to increase understanding of the general environment, firms engage in external environmental analysis. This analysis has four parts: scanning, monitoring, forecasting, and assessing. Analyzing the external environment is a difficult, yet significant, activity.

Identifying opportunities and threats is an important objective of studying the general environment. An opportunity is a condition in the general environment that, if exploited effectively, helps a company achieve strategic competitiveness. For example, recent market research results suggested to Procter & Gamble (P&G) after its acquisition of Gillette, a shaving products company, that an increasing number of men across the globe are interested in fragrances and skin care products. To take advantage of this opportunity, P&G is reorienting toward beauty products to better serve both men and women. The change constitutes an organization change focused on combining product categories rather than its typical organization around a specific branded product.

Environmental scanning provides reasonably hard data on the present situation and current trends, but intuition and luck are needed to accurately predict if these trends will continue. The resulting forecasts are, however, usually based on a set of assumptions that may or may not be valid. 环境扫描理性地提供了有关目前情况和当前趋势的坚实数据,但是还需要依靠直觉和运气以预测这些趋势是否会继续下去。

Various techniques are used to forecast future situations, and each has its proponents and critics. The most popular forecasting technique is extrapolation—the extension of present trends into the future. Trend extrapolation rests on the assumption that the world is reasonably consistent and changes slowly in the short run. Approaches of this type include time-series methods, which attempt to carry a series of historical events forward into the future. 最流行的预测方法是外推法——将目前的趋势扩展到未来。趋势外推法的假设是世界是同质的,并且在短期内变化缓慢。时间序列方法试图将一系列历史事件带入未来。

Scenarios are focused descriptions of different likely futures presented in a narrative fashion. 情境分析以叙事的方式重点描述各种不同的、可能的未来。

An industry scenario is a forecasted description of a particular industry's likely future. It is a scenario that is developed by analyzing the probable impact of future societal forces on key groups in a particular industry. 行业情境分析是预测特定行业未来可能性的文字描述。通过分析某个特定行业中未来各种社会驱动力量对关键群体可能产生的影响,撰写出行业情境分析报告。

A threat is a condition in the general environment that may hinder a company's efforts to achieve strategic competitiveness. Microsoft is currently experiencing a severe external threat as smartphones are expected to surpass personal computer (PC) sales in the near future. Although Microsoft has a smartphone operating system, Apple, Google, and Research in Motion (BlackBerry phones) have operating system, Apple, Google, and Research in motion (BlackBerry phones) have operating platforms that are much more popular than those using Microsoft's platform. Although PC growth will continue to expand, it is not growing at the rate that smartphones are, and possible substitution may happen between PCs, smartphones, and additional devices, such as Apple's iPad and similar devices. The main software platform is needed to assure other software producers will develop applications for the platform. Apple has large numbers of applications being developed, and Google's Android system software applications are rapidly increasing as well. As such, Microsoft is in a severe catch-up position relative to its competition. It recently formed a joint venture with Nokia Corporation to establish a firmer platform for its existing software using Nokia Corporation to establish a firmer platform for its existing software using Nokia's large potential smartphone base. However, this threat remains until this opportunity is realized.

Firms use several sources to analyze the general environment, including a wide variety of printed materials (such as trade publications, newspapers, business publications, and the results of academic research and public polls), trade shows and suppliers, customers, and employees of public-sector organizations. People in boundary-spanning positions can obtain a great deal of this type of information. Salespersons, purchasing managers, public relations directors, and customer service representatives, each of whom interacts with external constituents, are examples of boundary-spanning positions.

2.3   Scanning

Scanning entails the study of all segments in the general environment. Through scanning firms identify early signals of potential changes in the general environment and detect changes that are already under way. Scanning often reveals ambiguous, incomplete, or unconnected data and information. Thus, environmental scanning is challenging but critically important for firms, especially those competing in highly volatile environments. In addition, scanning activities must be aligned with the organizational context; a scanning system designed for a volatile environment is inappropriate for a firm in a stable environment.

Many firms use special software to help them identify events that are taking place in the environment and that are announced in public sources. For example, news event detection uses information-based systems to categorize text and reduce the trade-off between an important missed event and false alarm rates. The Internet provides significant opportunities for scanning. Amazon. com, for example, records significant information about individuals visiting its Web site, particularly if a purchase is made. Amazon then welcome these customers by name when they visit the Web site again. The firm sends messages to customers about specials and new products similar to those they purchased in previous visits. A number of other companies such as Netflix also collect demographic data about their customers in an attempt to identify their unique preferences(demographics is one of the segments in the general environment).

Philip Morris International continuously scans segments of its external environment to detect current conditions and to anticipate changes that might take place in different segments. For example, PMI always studies various nations' tax policies on cigarettes (these policies are part of the political/legal segment). The reason for this is that raising cigarette taxes might reduce sales while lowering these taxes might increase sales.

2.4　Industry Environment Analysis

An industry is a group of firms producing products that are close substitutes. In the course of competition, these firms influence one another. Typically, industries include a rich mixture of competitive strategies that companies use in pursuing above-average returns, In part, these strategies are chosen because of the influence of an industry's characteristics.

Compared with the general environment, the industry environment has a more direct effect on the firm's strategic competitiveness and ability to earn above-average returns. An industry's profit potential is a function of five forces of competition: the threats posed by new entrants, the power of suppliers, the power of buyers, product substitutes, and the intensity of rivalry among competitors(see Table 2.2).

Table 2.2　The Five Forces of Competition Model

Threat of new entrants
Bargaining power of suppliers
Bargaining power of buyers
Threat of substitute products
Rivalry among competing firms

The five forces model of competition expands the arena for competitive analysis. Historically, when studying the competitive environment, firms concentrated on companies with which they competed directly. However, firms must search more broadly to recognize current and potential competitors by identifying potential customers as well as the firms serving them. For example, the communications industry is now broadly defined as encompassing media companies, telecos, entertainment companies, and companies producing devices such as smartphones. In such an environment, firms must study many other industries to identify firms with capabilities (especially technology-based capabilities) that might be the foundation for producing a good or a service that can compete against what they are producing. Using this perspective finds firms focusing on customers and their needs rather than on specific industry boundaries to define markets.

When studying the industry environment, firms must also recognize that suppliers can become a firm's competitors (by integrating forward) as can buyers (by integrating backward). For example, several firms have integrated forward in the pharmaceutical industry by acquiring distributors or wholesalers. In addition, firms choosing to enter a new market and those producing products that are adequate substitutes for existing products can become a company's competitors. Next, we examine the five force the firm analyzes to understand the profitability potential within the industry (or a segment of an industry) in which it competes or may choose to compete.

Industry analysis refers to an in-depth examination of key factors within a corporation's task environment. 行业分析是指全面深入地检验公司任务环境中的关键驱动力量。

Rivalry is the amount of direct competition in an industry. 竞争对抗性是一个行业内直接竞争者的数量。

According to Porter, intensive rival is related to the presence of the following factors: number of competitors, rate of industry growth, product or service characteristics, amount of fixed costs, capacity, height of exit barriers, diversity of rivals. 根据波特理论，竞争的激烈程度与以下因素相关：竞争者的数量、行业的增长率、产品或服务特征、固定成本的多少、产能、退出壁垒的高度、竞争对手的多样性。

Suppliers can affect an industry through their ability to raise prices or reduce the quality of purchased goods and services. 供应商影响行业的能力可以通过提高价格或降低产品和服务的质量来实现。

2.5   Threat of New Entrants

Identifying new entrants is important because they can threaten the market share of existing competitors. One reason new entrants pose such a threat is that they bring additional production capacity. Unless the demand for a good or service is increasing, additional capacity holds consumers' costs down, resulting in less revenue and lower returns for competing firms. Often, new entrants have a keen interest in gaining a large market share. As a result, new competitors may force existing firms to be more efficient and to learn how to compete on new dimensions (e. g., sing an Internet-based distribution channel).

New entrants are newcomers to an existing industry. They typically bring new capacity, a desire to gain market share, and substantial resources. 新进入者是指现有行业的新参与者。它们通常带来新的产能、获得市场份额的渴望以及充足的资源。

An entry barrier is an obstruction that makes it difficult for a company to enter an industry. 进入壁垒是提高公司进入一个行业的困难程度的障碍。

Some of the possible barriers to entry involve economic of scale, product differentiation, capital requirements, switching costs, access to distribution channels, cost disadvantages independent of size, government policy and so on. 一些可能的进入壁垒包括规模经济、产品差异化、资本要求、转换成本、进入销售渠道、与成本无关的劣势、政府政策等。

The likelihood that firms will enter an industry is a function of two factors：barriers to entry and the retaliation expected from current industry is a function of two factors：barriers to entry and the retaliation expected from current industry participants. Entry barriers make it difficult for new firms to enter an industry and often place them at a competitive disadvantage even when they are able to enter. As such, high entry barriers tend to increase the returns for existing firms in the industry and may allow some firms to dominate the industry. Thus, firms competing successfully in an industry want to maintain high entry barriers in order to discourage potential competitors from deciding to enter the industry.

Bargaining Power of Suppliers

Increasing prices and reducing the quality of their products are potential means suppliers use to exert power over firms competing within an industry. If a firm is unable to recover cost increases by its suppliers through its own pricing structure, its profitability is reduced by its supplier's actions. A supplier group is powerful when

It is dominated by a few large companies and is more concentrated than the industry to which it sells.

Satisfactory substitute products are not available to industry firms.

Industry firms are not a significant customer for the supplier group.

Suppliers' goods are critical to buyers' marketplace success.

The effectiveness of suppliers' products has created high switching costs for industry firms.

It poses a credible threat to integrate forward into the buyers' industry. Credibility is enhanced when suppliers have substantial resources and provide a highly differentiated product.

The airline industry is one in which suppliers' bargaining power is changing. Though the number of suppliers is low, the demand for major aircraft is also relatively low. Boeing and Airbus aggressively compete for orders of major aircraft, creating more power for buyers in the process. When a large airline signals that it might place a "significant" order for wide-body airliners which either Airbus or Boeing might produce, both companies are likely to battle for the business and include a financing arrangement, highlighting the buyer's power in the potential transaction.

2.6   Bargaining Power of Buyers

Firms seek to maximize the return on their invested capital. Alternatively, buyers(customers of an industry or a firm) want to buy products at the lowest possible price-the point at which the industry earns the lowest acceptable rate of return on its invested capital. To reduce their costs, buyers bargain for higher quality, greater levels of service, and lower prices. These outcomes are achieved by encouraging competitive battles among the industry's firms. Customers(buyer groups) are powerful when

They purchase a large portion of an industry's total output.

The sales of the product being purchased account for a significant portion of the seller's annual revenues.

They could switch to another product at little, if any, cost.

The industry's products are undifferentiated or standardized, and the buyers pose a credible threat if they were to integrate backward into the sellers' industry.

Consumers armed with greater amounts of information about the manufacturer's costs and the power of the Internet as a shopping and distribution alternative have increased bargaining power in many industries. One reason for this shift is that individual buyers incur virtually zero switching costs when they decide to purchase from one manufacturer rather than another or from one dealer as opposed to any other.

Buyers affect an industry through their ability to force down prices, bargain for higher quality or more services, and play competitors against each other. 购买者影响行业的能力通过以下几个方面实现:强迫降价,对交易提供更高的质量或更多的服务讨价还价,以及发挥竞争者的作用。

2.7   Threat of Substitute Products

Substitute products are goods or service from outside a given industry that perform similar or the same functions as a product that the industry produces. For example, as a sugar substitute, NutraSweet (and other sugar substitutes) place an upper limit on sugar manufacturers' prices—NutraSweet and sugar perform the same function, though with different characteristics. Other product substitutes include e-mail and fax machines instead of overnight deliveries, plastic containers rather than glass jars, and tea instead of coffee. Newspaper firms have experienced significant circulation declines over the past decade or more. The declines are due to substitute outlets for news including Internet sources, cable television news channels, and e-mail and cell phone alerts. Likewise, satellite TV and cable and telecommunication companies provide substitute services for basic media services such as television, Internet, and phone. However, as illustrated in the Strategic Focus, the possible switching is becoming more complicated as consumer demand for content changes through increasing use of mobile devices such as tablets and smartphones. Tablets such as the iPad are reducing the number of PCs sold and this is curtailing the growth of PC producers such as China Taiwan's Acer Computers, at least until they can come out with their own successful tablet product. These products are increasingly popular, especially among younger and technologically savvy people, and as product substitutes they have significant potential to continue to reduce traditional media sources such as newspaper circulation sales.

In general, product substitutes present a strong threat to a firm when customers face few, if any, switching costs and when the substitute product's price is lower or its quality and performance capabilities are equal to or greater than those of the competing product. Differentiating a product along dimensions that customers value (such as quality, service after the sale, and location) reduces a substitute's attractiveness.

Substitute products are those products that appear to be different but can satisfy the same need as another product. 替代品是指那些看似不同但能满足相同需要的其他产品。

According to Porter, "substitutes limit the potential returns of industry by placing a ceiling on the prices firms in the industry can profitably charge."根据波特理论,一个产业的潜在收益之所以是有限的是由于替代品为这个行业设置了价格上限。

A sixth force should be added to Porter's list to include a variety of stakeholder groups from the task environment. 应该加入波特五力清单的第六种力量,包括来自任务环境的各种利益相关者群体。

A complementor is a company (e. g. , Microsoft) or an industry whose product works well with a firm's (e. g. , Intel's) product and without which the product would lose much of its value.互补者是指一个公司(如微软)或行业的产品与其他公司(如英特尔)的产品结合将发挥更大的作用,否则产品将失去很多优良价值。

2.8   Strategic Groups

A set of firms that emphasize similar strategic dimensions and use a similar strategy is called a strategic group. The competition between firms within a strategic group is greater than the competition between a member of a strategic group and companies outside that strategic group. Therefore, intrastrategic group competition is more intense than is interstragegic group competition. In fact, more heterogeneity is evident in the performance of firms within strategic groups than across the groups. The performance leaders within groups are able to follow strategies similar to those of other firms in the group and yet maintain strategic distinctiveness to gain and sustain a competitive advantage.

The extent of technological leadership, product quality, pricing policies, distribution channels, and customer service are examples of strategic dimensions that firms in a strategic group may treat similarly. Thus, membership in a particular strategic group defines the essential characteristics of the firm's strategy.

The notion of strategic groups can be useful for analyzing an industry's competitive structure. Such analyses can be helpful in diagnosing competition, positioning, and the profitability of firms within an industry. High mobility barriers, high rivalry, and low resources among the firms within an industry limit the formation of strategic groups. However, research suggests that after strategic groups are formed, their membership remains relatively stable over time, although recent research does examine how change occurs. Using strategic groups to understand an industry's competitive structure requires the firm to plot companies' competitive actions and competitive responses along strategic dimensions such as pricing decisions, product quality, distribution channels, and so forth. This type of analysis shows the firm how certain companies are competing similarly in terms of how they use similar strategic dimensions.

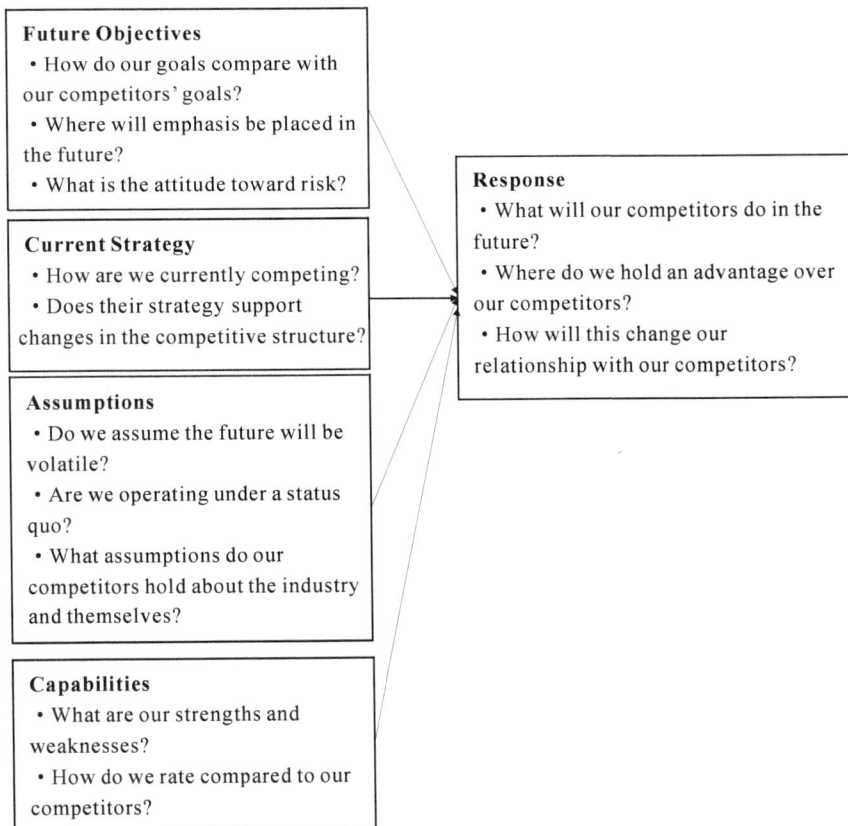

Future Objectives
- How do our goals compare with our competitors' goals?
- Where will emphasis be placed in the future?
- What is the attitude toward risk?

Current Strategy
- How are we currently competing?
- Does their strategy support changes in the competitive structure?

Assumptions
- Do we assume the future will be volatile?
- Are we operating under a status quo?
- What assumptions do our competitors hold about the industry and themselves?

Capabilities
- What are our strengths and weaknesses?
- How do we rate compared to our competitors?

Response
- What will our competitors do in the future?
- Where do we hold an advantage over our competitors?
- How will this change our relationship with our competitors?

Figure 2.3　Competitor Analysis Components

35

Multinational corporation (MNC) is a company having significant manufacturing and marketing operations in multiple countries. 跨国公司是指在多个国家从事大量生产和营销业务的公司。

Before a company plans its strategy for a particular international location, it must scan the particular country's societal environment in question for opportunities and threats and compare them to its own organizational strengths and weaknesses. 一个公司在为特定国际市场制定战略计划之前，必须认真扫描特定国家的社会环境，发现机会和威胁，并与组织自身的优势和劣势相比较。

Strategic groups have several implications. First, because firms within a group offer similar products to the same customers, the competitive rivalry among them can be intense. The more intense the rivalry, the greater the threat to each firm's profitability. Second, the strengths of the ve industry forces differ across strategic groups. Third, the closer the strategic groups are in terms of their strategies, the greater is the likelihood of rivalry between the groups.

2.9　Ethical Considerations

Firms must follow relevant laws and regulations as well as carefully articulated ethical guidelines when gathering competitor intelligence. Industry associations often develop lists of these practices that firms can adopt. Practices considered both legal and ethical include (1) obtaining publicly available information (e. g. , court records, competitors' help-wanted advertisements, annual reports of publicly held corporations, and Uniform Commercial Code filings), and(2) attending trade fairs and shows to obtain competitors' brochures, view their exhibits, and listen to discussions about their products. In contrast, certain practices(including blackmail, trespassing, eavesdropping, and stealing drawings, samples, or documents) are widely viewed as unethical and often are illegal.

The industry life cycle is useful for explaining and predicting trends among the six forces that drive industry competition. 行业生命周期可以用来解释和预测驱动行业竞争的六种驱动力量的趋势。

By the time an industry enters maturity, products tend to become more like commodities. This is now a consolidated industry—dominated by a few large firms, each of which struggles to differentiate its products from the competitors. 当一个行业进入成熟期，产品出现变为普通商品的趋势。如今行业已经成熟——几家大公司主导行业，彼此通过产品差异化相互竞争。

No firm has large market share and each firm serves only a small piece of the total market in competition with others in a fragmented industry. 在分散行业中，没有任何企业占有较高的市场份额，每个企业在总的市场竞争中只占据很小一部分。

36

As an industry moves through maturity toward possible decline, the growth rate of its products' sales slows and may even begin to decrease. 当行业从成熟阶段转向衰退阶段,产品销售的增长速度减缓,甚至可能开始下降。

A global industry, in contrast, operates world-side, with MNCs making only small adjustments for country-specific circumstances. 全球性行业在世界各地经营运作,跨国公司根据特定国家的具体情况所做出的调整非常小。

Some competitor intelligence practices may be legal, but a firm must decide whether they are also ethical, given the image it desires as a corporate citizen. Especially with electronic transmissions, the line between legal and ethical practices can be difficult to determine. For example, a firm may develop Web site addresses that are similar to those of its competitors and thus occasionally receive e-mail transmissions that were intended for those competitors. The practice is an example of the challenges companies face in deciding how to gather intelligence about competitors while simultaneously determining how to prevent competitors from learning too much about them. To deal with these challengers, firms should establish principles and take actions that are consistent with them. Many firms follow the Strategy and Competitive Intelligence Professionals, a professional association, code of professional practice and ethics dealing with this issue.

Open discussions of intelligence-gathering techniques can help a firm ensure that employees, customers, suppliers, and even potential competitors understand its convictions to follow ethical practices for gathering competitor intelligence. An appropriate guideline for competitor intelligence practices is to respect the principles of common morality and the right of competitors not to reveal certain information about their products, operations, and strategic intentions.

Using an EFAS (External Factors Analysis Summary) Table is one way to organize the external factors into the generally accepted categories of opportunities and threats as well as to analyze how we a particular company's management(rating) is responding to these specific factors in light of the perceived importance(weight) these factors the company.

Table 2.3   External Factor Analysis Summary Table for Maytag

External Factors	Weight	Rating	Weighted Score	Comments
Opportunities				
• Economic integration of European Union	0.20	4	0.80	Acquisition of Hoover
• Demoqraphics favor quality appliances	0.10	5	0.50	Maytag quality
• Economic development of Asia	0.05	1	0.05	Low Maytag presence
• Opening of Eastern Europe	0.05	2	0.10	Will take time

37

External Factors	Weight	Rating	Weighted Score	Comments
• Trend to superstores	0.10	2	0.20	Maytag weak in this channel
Threats				
• Increasing government regulations	0.10	4	0.40	Well positioned
• Strong U. S. competition	0.10	4	0.40	Well positioned
• Whirlpool and Electrolux strong globally	0.15	3	0.45	Hoover weak globally
• New product advances	0.05	1	0.05	Questionable
• Japanese appliance companies	0.10	2	0.20	Only Asian presence is Australia
Totals	1.00		3.15	

2.10   Rivalry Among Competing Sellers

Usually the most powerful of the five forces. The big factor determining the strength of rivalry is how actively and aggressively are rivals employing the various weapons of competition in jockeying for a stronger market position and seeking bigger sales Is price competition vigorous? Active efforts to improve quality? Are rivals racing to offer better performance features? Are rivals racing to offer better customer service? Lots of advertising/sales promotions? Active efforts to build a stronger dealer network? Active product innovation? Active use of other weapons of rivalry?

What Causes Rivalry to be Stronger? Active jockeying for position among rivals and frequent launches of new offensives to gain sales and market share. One or more firms initiates moves to bolster their standing at expense of rivals. Lots of firms that are relatively equal in size and capability. Slow market growth Industry conditions tempt some firms to go on the offensive to boost volume and market share. Customers have low costs in switching to rival brands. A successful strategic move carries a big payoff Costs more to get out of business than to stay in Firms have diverse strategies, corporate priorities, resources, and countries of origin.

Principle of Competitive Markets Competitive jockeying among rival firms is dynamic and ever-changing. As industry members initiate new offensive and defensive moves. As emphasis swings from one mix of competitive weapons to another.

Competitive Force of Potential Entry Seriousness of threat depends on. Barriers to entry. Reaction of existing firms to entry. Barriers exist when Newcomers confront obstacles Economic factors put potential entrant at a disadvantage relative to incumbent firms.

Common Barriers to Entry Sizable economies of scale In ability to gain access to specialized technology Existence of strong learning/experience curve effects. Strong brand preferences and customer loyalty. Large capital requirements and/or other specialized resource requirements. Cost disadvantages independent of size Difficulties in gaining access to distribution channels Regulatory policies，tariffs，trade restrictions.

Principle of Competitive Markets Threat of entry is stronger when：Entry barriers are low Sizable pool of entry candidates exists Incumbents are unwilling or unable to contest a newcomer's entry efforts. Newcomers can expect to earn attractive profits.

Competitive Force of Substitute Products Concept Substitutes matter when customers are attracted to the products of firms in other industries Eyeglasses vs. Contact Lens Sugar vs. Artificial Sweeteners Newspapers vs. TV vs. Internet E-mail vs. Overnight Delivery Examples.

How to Tell Whether Substitute Products are a Strong Force Sales of substitutes are growing rapidly Producers of substitutes plan to add new capacity Profits of producers of substitutes are up.

Principle of Competitive Markets Competitive threat of substitutes is stronger when they are：Readily available Attractively priced Believed to have comparable or better performance features Customer switching costs are low.

A strategic group is a set of business units or firms that "pursue similar strategies with similar resources. " 战略集团是以"类似资源寻求类似战略"的一系列业务单位或企业。

In analyzing the level of competitive intensity within a particular industry or strategic group，it is useful to characterize the various competitors for predictive purposes. A strategic type is category of firms based on a common strategic orientation and a combination of structure，culture，and processes consistent with that strategy. According to Miles and Snow，competing firms within a single industry can be categorized on the basis of their general strategic orientation into one of four basic types：defenders，prospectors，analyzers，and reactors. 战略类型是基于共同的战略导向，并综合考虑结构、文化和流程与战略的一致性，对企业进行分类的方法。根据迈尔斯和斯诺的研究，单个行业内的竞争性公司可以基于总体战略导向分为四种基本类型：防御者、探索者、分析者和反应者。

Defenders are companies with a limited product line that focus on improving the efficiency of their existing operations. 防御者是那些在有限的产品线中关注于提高目前的运行效率的公司。

Prospectors are companies with fairly broad product lines that focus on product innovation and market opportunities. 探索者是指有相当广泛的产品线,专注于产品创新和市场机会的公司。

Analyzers are companies that operate in at least two different product-market areas, one stable and one variable. 分析者是指那些至少在两个不同的产品市场领域运作,一个是稳定的市场,另一个是多样化的市场。

Reactors are companies that lack a consistent strategy-structure relationship. 反应者是指那些与战略、结构、文化缺乏一致联系的公司。

Competitive Pressures From Suppliers and Supplier-Seller Collaboration. Whether supplier-seller relationships represent a weak or strong competitive force depends on Whether suppliers can exercise sufficient bargaining leverage to influence terms of supply in their favor Extent and competitive importance of collaborative partnerships between one or more sellers and their suppliers.

Competitive Force of suppliers are a strong competitive force when: Item makes up large portion of product costs, is crucial to production process, and/or significantly affects product quality. It is costly for buyers to switch suppliers. They have good reputations and growing demand. They can supply a component cheaper than industry members can make it themselves. They do not have to contend with substitutes Buying firms are not important customers.

Competitive Pressures: Collaboration Between Sellers and Suppliers Rival sellers are forming long-term strategic partnerships with select suppliers to Promote just-in-time deliveries and reduced inventory and logistic costs. Speed availability of next-generation components. Enhance quality of parts being supplied Reduce suppliers' costs which paves way for lower prices on items supplied. Competitive advantage potential may accrue to industry rivals doing the best job of managing supply-chain relationships.

Principle of Competitive Markets Suppliers are a stronger force the more they can exercise power over: Prices charged Quality and performance of items supplied Reliability of deliveries.

Competitive Pressures From Buyers and Seller-Buyer Collaboration. Whether seller-buyer relationships represent a weak or strong competitive force depends on. Whether buyers have sufficient bargaining leverage to influence terms of sale in their favor Extent and competitive importance of collaborative partnerships between one or more sellers and their customers.

Hypercompetition describes an industry undergoing an ever-increasing level of environmental uncertainty in which competitive advantage is only temporary. 超级竞争描述了一个行业正在经历环境不确定程度不断提高的形势。

Competitive intelligence is a formal program of gathering information on a company's competitors. Sometimes called business intelligence, this is one of the fastest growing fields in strategic management. 竞争情报是一个收集公司竞争对手信息的正式项目。有时也称为商业情报,是战略管理增长最快的领域之一。

2.11   Control and Environmental Variables

In the basic regression that we will consider, the control and environmental variables are a measure of international openness, the ratio of government consumption to GDP, a subjective indicator of maintenance of the rule of law, a subjective indicator of democracy (electoral rights), the log of the total fertility rate, the ratio of real gross domestic investment to real GDP, and the inflation rate. The system also includes the contemporaneous growth rate of the terms of trade, interacted with the extent of international openness (the ratio of exports plus imports to GDP). We take account of the likely endogeneity of the explanatory variables by using lagged values as instruments. These lagged variables may be satisfactory because the error term in the equation for the per capita growth rate turns out to display little serial correlation.

In the neoclassical growth models of Solow – Swan and Ramsey, the effects of the control and environmental variables on the growth rate correspond to their influences on the steady-state position. For example, an exogenously higher value of the rule of law indicator raises the steady-state level of output per effective worker. The growth rate, Dyt , tends accordingly to increase for given values of the state variables. Similarly, a higher ratio of (nonproductive) government consumption to GDP tends to depress the steady-state level of output per effective worker and thereby reduce the growth rate for given values of the state variables.

In neoclassical growth models, a change in a control or environmental variable affects the steady-state level of output per effective worker, but not the long-term per capita growth rate. The long-run or steady-state growth rate is given by the rate of exogenous technological progress. In contrast, in the endogenous-growth models, variables that affect R&D intensity also influence long-term growth rates. However, even in the Solow—Swan and Ramsey models, if the adjustment to the new steady-state position takes a long time—as seems to be true empirically—then the growth effect of a variable such as the rule-of-law indicator or the government consumption ratio lasts for a long time.

The measures of educational attainment that we use in the main analysis are based on years of schooling and do not adjust for variations in school quality. A measure of quality, based on internationally comparable test scores, turns out to have much more explanatory power for growth. However, this test-score measure is unavailable for much of the sample and is, therefore, excluded from the basic system.

Health capital is proxies in the basic system by the reciprocal of life expectancy at age one. If the probability of dying were independent of age, then this reciprocal would give the probability per year of dying. We also consider later measures of infant mortality and child mortality, as well as incidence of a specific disease, malaria. We assume that the government consumption variable measures expenditures that do not directly affect productivity but that entail distortions of private decisions. These distortions can reflect the governmental activities themselves and also involve the adverse effects from the associated public finance. A higher value of the government consumption ratio leads to a lower steady-state level of output per effective worker and, hence, to a lower growth rate for given values of the state variables.

The fertility rate is an important influence on population growth, which has a negative effect on the steady-state ratio of capital to effective worker in the neoclassical growth model. Hence, we anticipate a negative effect of the fertility rate on economic growth. Higher fertility also reflects greater resources devoted to child rearing, as in the model developed . This channel provides another reason why higher fertility would be expected to reduce growth. The effect of the saving rate in the neoclassical growth model is measured empirically by the ratio of real investment to real GDP. Recall that we attempt to isolate the effect of the saving rate on growth, rather than the reverse, by using lagged values—in this case, the lagged investment ratio—as instruments.

We assume that an improvement in the rule of law, as gauged by the subjective indicator provided by an international consulting firm (Political Risk Services), implies enhanced property rights and, therefore, an incentive for higher investment and growth. More broadly, the idea is that well-functioning political and legal institutions help to sustain growth. Some historical analyses attempt to relate current institutional characteristics, such as maintenance of the rule of law, to practices of colonial powers long ago. Robinson argue that European colonists were more likely to invest in institutions in regions that were previously poor or empty, notably present-day Canada and the United States, because they lacked the potential for exploitation of mineral wealth and indigenous populations. Robinson stress that the adverse mortality experience of settlers in parts of Latin America and Africa may have limited institutional investments in those colonies. Woodberry (2002) argues that the establishment of quality schooling by missionaries in some colonies may have had a long-lasting influence on political institutions. These analyses suggest instrumental variables—from the long-term history—that can be used to get more reliable estimates of the effects of current variables, such as the rule-of-law indicator.

We also include another subjective indicator (from Freedom House) of the extent of democracy in the sense of electoral rights. Theoretically, the effect of democracy on growth is ambiguous. Negative effects arise in political models that stress the incentive of electoral majorities to use their political power to transfer resources away from rich minority groups. Democracy may also be productive as a mechanism for government to commit itself not to confiscate the capital accumulated by the private sector. The empirical analysis allows for a linear and squared term in democracy, thereby allowing for the possibility that the sign of the net effect would depend on the extent of democracy.

The explanatory variables also include a measure of the extent of international openness— the ratio of exports plus imports to GDP. Openness is well known to vary by country size—larger countries tend to be less open because internal trade offers a large market that can substitute effectively for international trade. The explanatory variable used in the analysis of growth filters out the normal relationship (estimated in another regression system) of international openness to the logs of population and area. This filtered variable reflects especially the influences of government policies, such as tariffs and trade restrictions, on international trade.

We include also the growth rate over each decade of the terms of trade, measured by the ratio of export prices to import prices. This ratio appears as a product with the extent of openness, measured by exports plus imports over GDP. This terms-of-trade variable measures the effect of changes in international prices on the income position of domestic residents. This real income position would rise because of higher export prices and fall with higher import prices. We view the terms of trade as determined on world markets and, hence, exogenously to the behavior of an individual country.

Since an improvement in the terms of trade raises a country's real income, we would predict an increase in domestic consumption. An effect on production, GDP, depends, however, on a response of allocations or effort to the shift in relative prices. If an increase in the relative price of the goods that a country produces tends to generate more output, that is, a positive response of supply, then the effect of this variable on the growth rate would be positive. One effect of this type is that an increase in the relative price of oil—an import for most countries—would reduce the production of goods that use oil as an input. Finally, the basic system includes the average inflation rate as a measure of macroeconomic stability. Alternative measures can also be considered, including fiscal variables.

Chapter 3 Internal Environment Analysis

3.1 Analyzing the Internal Organization

One of the conditions associated with analyzing a firm's internal organization is the reality that in today's global economy, some of the resources that were traditionally critical to firm's efforts to produce, sell, and distribute their goods or services such as labor costs, access to financial resources and raw materials, and protected or regulated markets are still important; but, it is now less likely that these resources will become core competencies and possibly competitive advantages. An important reason for this is that an increasing number of firms are using their resources to form core competencies through which they successfully implement an international strategy as a means of overcoming the advantages created by these more traditional resources.

The Volkswagen Group has established "Strategy 2018" as its international strategy. The firm, which sells its products in over 150 countries, employs 400,000 people to operate its 62 production plants located in 15 European countries. By using its resources to form technological and innovation capabilities, Volkswagen intends to create superior customer service and quality as core competencies on which it will rely to implement its international strategy.

Organizational analysis is concerned with identifying and developing an organization's resources. 内部扫描通常称为组织分析,主要考虑识别和开发组织资源的有关问题。

Resources are organization's assets and are thus its basic building blocks. They include tangible assets such as plant, equipment, finances, and location; human assets, in terms of the number of employees and their skills; and intangible assets, such as technology, culture, and reputation. 资源是一个组织的资产,因而属于基本必备条件。有形资源如厂房、设备、资金和位置;人力资源是指一定数量的员工及其技能;无形资源包括技术、文化和声誉。

3.2 Creating Value

Firm use their resources as the foundation for producing goods or services that will create value for customers. Value is measured by a product's performance characteristics and by its attributes for which customers are willing to pay. Firms create value by innovatively bundling and leveraging their resources to form capabilities and core competencies. Firms with a competitive advantage create more value for customers than do competitors. Walmart uses its "every day low price" approach to doing business (an approach that is grounded in the firm's core competencies, such as information technology and distribution channels) to create value for those seeking to buy products at a low price compared to competitors' price for those seeking to buy products at a low price compared to competitors' price for those products. Mattress manufacture E. S. Kluft. (The firm's upper-end mattress sells for $50000 per unit.) The stronger these firms' core competencies, the grater the amount of value they're able to create for their customers.

Ultimately, creating value for customers is the source of above-average returns for a firm. What the firm intends regarding value creation affects its choice of business-level strategy and its organizational structure. In discussion of business-level strategies, we note that value is created by a product's low cost, by its highly differentiated features, or by a combination of low cost and high differentiation, compared with competitors' offerings. A business-level strategy is effective only when it is grounded in exploiting the firm's capabilities and core competencies. Thus, the successful firm continuously examines the effectiveness of current capabilities and core competencies while thinking about the capabilities and competencies it will require for future success.

Capabilities refer to a corporation's ability to exploit its resources. 能力是指一个公司运用资源的能力。

A competency is the cross-functional integration and coordination of capabilities. 竞争力是指跨职能部门进行整合和协调的能力。

A core competency is a collection of competencies that cross divisional boundaries, is widespread within the corporation, and is something that a corporation can do exceedingly well. 核心竞争力是公司内部普遍存在的各种跨部门边界竞争力的集合, 也是一些企业表现出色的原因所在。

A core rigidity is a strength that over time matures and becomes a weakness. 所谓核心刚性是指某种优势随着时间的推移逐渐成熟并成为劣势。

When core competencies are superior to those of the competition, they are called distinctive competencies. 当某种核心竞争力优于竞争对手相应的能力时, 则称为独特竞争力。

At one time, the firm's efforts to create value were largely oriented to understanding the characteristics of the industry in which it competed and, in light of those characteristics, determining how it should be positioned relative to competitors. This emphasis on industry characteristics and competitive strategy underestimated the role of the firms' resources and capabilities in developing core competencies as the source of competitive advantages. In fact, core competencies, in combination with product-market positions, are the firm's most important sources of competitive advantage. A firm's core competencies, integrated with an understanding of the results of studying the conditions in the external environment, should drive the selection of strategies. As Clayton Christensen noted, "Successful strategists need to cultivate a deep understanding of the processes of competition and progress and of the factors that undergird each advantage. Only thus will they be able to see when old advantages are poised to disappear and how new advantages can be built in their stead." By emphasizing learn to competencies when selecting and implementing strategies, companies learn to compete primarily on the basis of firm-specific differences. However, while doing so they must be simultaneously aware of how things are changing in the external environment.

Table 3.1 Conditions Affecting Managerial Decisions about Resources, Capabilities and Core Competencies

Conditions	Uncertainty	Uncertainty exists about the characteristics of the firm's general and industry environments and customers' needs
	Complexity	Complexity results from the interrelationships among conditions shaping a firm.
	Intra-organizational Conflicts	Intra-organizational conflicts may exist among managers making decisions as well as among those affected by the decisions.

3.3 Resources, Capabilities, and Core Competencies

Resources, capabilities, and core competencies are the foundation of competitive advantage. Resources are bundled to created organizational capabilities. In turn, capabilities are the source of a firm's core competencies, which are the basis of establishing competitive advantages. We show these relationships in Figure 3.2. Here, we define and provide examples of these building blocks of competitive advantage.

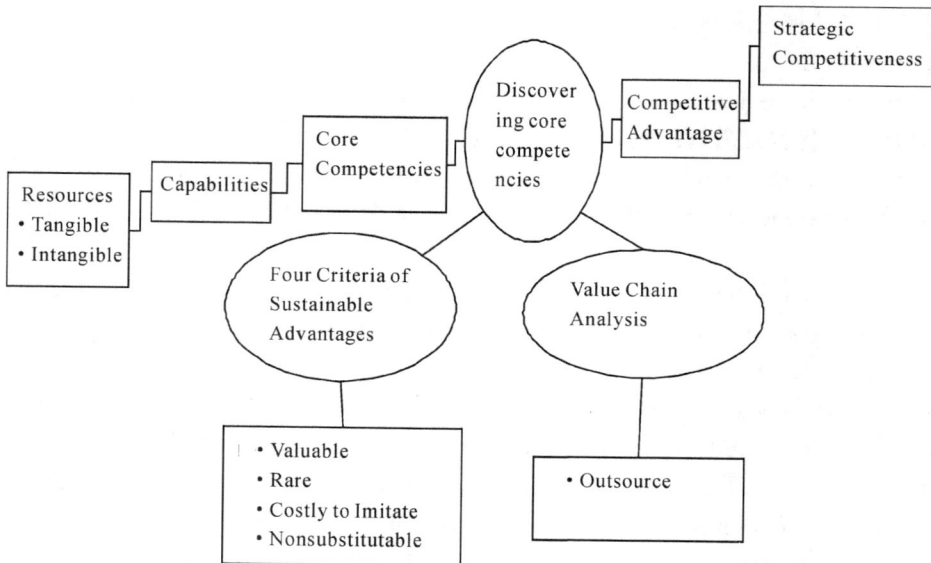

Figure 3.1 Components of an Internal Analysis

3.4   Resources

Broad in scope，resources over a spectrum of individual，social，and organizational phenomena. By themselves，resources do not allow firms to create value for customers as the foundation for earning above-average returns. Indeed，resources are combined to form capabilities. Subway links its fresh ingredients with several other resources including the continuous training it provides to those running the firm's units as the foundation for customer service as a capability；as explained in the Opening Case，customer service is also a core competence for Subway. As its sole distribution channel，the Internet is a resource for Amazon.com. The firm uses the Internet to sell goods at prices that typically are lower than those offered by competitors selling the same goods through what are more costly brick-and-mortar storefronts. By combining other resources（such as access to a wide product inventory），Amazon has developed a reputation for excellent customer service. Amazon's capability in terms of customer service is a core competence as well in that the firm creates unique value for customers through the services it provides to them. Amazon also uses its technological core competence to offer AWS（Amazon Web Services），services through which businesses can rent computing power from Amazon at a cost of pennies per hour. In the words of the leader of this effort，"AWS makes it possible for anyone with an Internet connection and a credit card to access the same kind of world-class computing systems that Amazon uses to run its $34 billion-a - year retail operation."

Proposing that a company's sustained competitive advantage is primarily determined by its resource endowments，Grant presents a five-step，resource-based approach to strategy analysis. 格兰特认为,公司可持续的竞争优势主要取决于自身的资源禀赋,并提出了以资源为基础的战略分析的五个步骤。

The sustainability of an advantage determines that durability and imitability. 优势的可持续性取决于持久性和模仿性。

Durability is the rate at which a firm's underlying resources，capabilities，or core competencies depreciate or become obsolete. 持久性是指公司的基础性资源、能力和核心竞争力贬值或过时的速度。

Imitability is the rate at which a firm's underlying resources，capabilities，or core competencies can be duplicated by others. 模仿性是指公司的基础性资源、能力和核心竞争力可以被别人复制的速度。

3.5   Tangible Resources

As tangible resources，a firm's borrowing capacity and the status of its physical

47

facilities are visible. The value of many tangible resources can be established through financial statements, but these statements do not account for the value of all the firm's assets, because they disregard some intangible resources. The value of tangible resources is also constrained because they are hard to leverage-it is difficult to derive additional business or value from a tangible resource. For example, an airplane is a tangible resource, but "You can't use the same airplane on five different routes at the same time. You can't put the same crew on five different routes at the same time". And the same goes for the financial investment you've made in the airplane.

Table 3.2 Tangible Resources

Financial Resources	• The firm's capacity to borrow • The firm's ability to generate funds through internal operations
Organizational Resources	• Formal reporting structures
Physical Resources	• The sophistication of a firm's plant and equipment and the attractiveness of its location • Distribution facilities • Product inventory
Technological Resources	• Availability of technology-related resources such as copyrights, patents, trademarks, and trade secrets

Table 3.3 Intangible Resources

Human Resources	• Knowledge • Trust • Skills • Abilities to collaborate with others
Innovation Resources	• Ideas • Scientific capabilities • Capacity to innovate
Reputational Resources	• Brand name • Perceptions of product quality, durability, and reliability • Positive reputation with stakeholders such as suppliers and customers

3.6 Intangible Resources

Compared to tangible resources, intangible resources are a superior source of capabilities and subsequently, core competencies. In fact, in the global economy, "the success of a corporation lies more in its intellectual and systems capabilities than in its physical assets. Moreover, the capacity to manage human intellect-and to convert it into useful products and services-is fast becoming the critical executive skill of the age".

Because intangible resources are less visible and more difficult for competitors to understand, purchase, imitate, or substitute for, firms prefer to rely on them rather than on tangible resources as the foundation for their capabilities. In fact, the more unobservable (i. e. , intangible) a resource is, the more valuable that resource is to create capabilities. Another benefit of intangible resources is that, unlike most tangible resources, their use can be lever-aged. For instance, sharing knowledge among employees does not diminish its value for any one person. To the contrary, two people sharing their individualized knowledge sets often can be leveraged to create additional knowledge that, although new to each individual, contributes to performance improvements for the firm.

Reputational resources are important sources of a firm's capabilities and core competencies. Indeed, some argue that a positive reputation can even be a source of competitive advantage. Earned through the firm's actions as well as its words, a value-creating reputation is a product of years of superior marketplace competence as perceived by stakeholders. A reputation indicates the level of awareness a firm has been able to develop among stakeholders and the degree to which they hold the firm in high esteem.

A well-known and highly valued brand name is a specific reputational resource. A continuing commitment to innovation and aggressive advertising facilitates firms' efforts to take advantage of the reputation associated with their brands. Harley-Davidson has a reputation for producing and servicing high-quality motorcycles with unique designs. Because of the desirability of its reputation, the company also produces a wide range of accessory items that it sells on the basis of its reputation for offering unique products with high quality. Sunglasses, jewelry, belts, wallets, shirts, slacks, belts, and hats are just a few of the large variety of accessories customers can purchase from a Harley-Davidson dealer or from its online store.

Generally speaking, each organization structure tend to support some corporate strategies over others. 一般来说,每一种组织结构都会倾向于支持某些企业战略。

The typical organization structure involves that simple structure, functional structure and divisional structure. 典型的组织结构包括简单结构、职能结构、分部结构。

Corporate culture is the collection of beliefs, expectations, and values learned and shared by a corporation's members and transmitted from one generation of employees to another. 企业文化是公司成员学习、共享并且代代相传的信念、期望和价值观的集合。

Cultural intensity (or depth) is the degree to which members of a unit accept the norms, values, or other culture content associated with the unit. 文化的强度或深度是指全体成员作为一个整体接受各种规范、价值观或其他相关文化内涵的程度。

Cultural integration (or breadth) is the extent to which units throughout an organization share a common culture. 文化的集成度或宽度是整个组织的各个单位共享文

化的程度。

3.7 Capabilities

The firm combines individual tangible and intangible resources to create capabilities. In turn, capabilities are used to complete the organizational tasks required to produce, distribute, and service the goods or services the firm provides to customers for the purpose of creating value for them. As a foundation for building core competencies and hopefully competitive advantages, capabilities are often based on developing, carrying, and exchanging information and knowledge through the firm's human capital. Hence, the value of human capital in developing and using capabilities and, ultimately, core competencies cannot be overstated. At IBM, for example, human capital is critical to forming and using the firm's capabilities for long-term customer relationships and deep scientific and research skills, and the breadth of the firm's technical skills, and the breadth of the firm's technical skills in hardware software, and services.

As illustrated in Table 3.3, capabilities are often developed in specific functional areas (such as manufacturing, R&D, and marketing) or in a part of a functional area(e. g., advertising). Table 3.3 shows a grouping of organizational functions and the capabilities that some companies are thought to possess in terms of all or parts of those functions.

Table 3.4　example of Firms' Capabilities

Functional Areas	Capabilities	Examples of Firms
Distribution	• Effective use of logistics management techniques	• Walmart
Human Resources	• Motivating, empowering, and retaining employees	• Microsoft
Management Information Systems	• Effective and efficient control of inventories through point-of-purchase data collection methods	• Walmart
Marketing	• Effective promotion of brand-name products • Effective customer service • Innovative merchandising	• Procter & Gamble • Ralph Lauren Corp. • Mckinsey & Co. • Nordstrom Inc. • Crate & Barrel
Management	• Ability to envision the future of clothing	• Hugo Boss • Zara

Manufacturing	• Design and production skills yielding reliable products • Product and design quality • Miniaturization of components and products	• Komatsu • Witt Gas Technology • Sony
Research & Development	• Innovative technology • Development of sophisticated elevator control solutions • Rapid transformation of technology into new products and processes • Digital technology	• Caterpillar • Otis Elevator Co. • Chaparral Steel • Thomson Consumer Electronics

3.8   Core Competencies

Defined in chapter, core competencies are capabilities that serve as a source of competitive advantage for a firm over its rivals. Core competencies distinguish a company competitively and reflect its personality. Core competencies emerge over time through an organizational process of accumulating and learning how to deploy different resources and capabilities. As the capacity to take action, core competencies are "crown jewels of a company," the activities the company performs especially well compared to competitors and through which the firm adds unique value to the goods or services it sells to customers.

Two tools help firms identify their core competencies. The first consists of four specific criteria of sustainable competitive advantage that can be used to determine which capabilities are core competencies. Because the capabilities shown in Table 3.3 have satisfied these four criteria, they are core competencies. The second tool is the value chain analysis. Firms use this tool to select the value-creating competencies that should be maintained, upgraded, or developed and those that should be outsourced.

3.9   The Four Criteria of Sustainable Competitive Advantage

Capabilities that are valuable, rare, costly to imitate, and nonsubstitutable are core competencies (see Table 3.4). In turn, core competencies can lead to competitive advantages for the firm over its rivals. Capabilities failing to satisfy the four criteria are not core competencies, meaning that although every core competence is a capability, not every capability is a core competence. In slightly different words, for a capability to be a core competence, it must be valuable and unique from a customer's point of view. For a core competence to be a potential source of competitive advantage, it must be inimitable and nonsubstitutable by competitors.

Table 3. 5   The Four Criteria of Sustainable Competitive Advantage

Valuable Capabilities	• Help a firm neutralize threats of exploit opportunities
Rare Capabilities	• Are not possessed by many others
Costly-to Imitate Capabilities	• Historical：A unique and a valuable organizational culture or brand name • Ambiguous cause：The causes and uses of a competence are unclear • Social complexity：Interpersonal relationships，trust，and friendship among managers，suppliers，and customers
Nonsubstituable Capabilities	• No strategic equivalent

A sustainable competitive advantage exists only when competitors cannot duplicate the benefits of a firm's strategy or when they lack the resources to attempt imitation. For some period of time, the firm may have a core competence by using capabilities that are valuable and rare, but imitable. For example, some firms are trying to develop a core competence and potentially a competitive advantage by out-greening their competitors. Since 2005，Walmart has used its resources in ways that have allowed it to reduce its stores' carbon footprint by more than 10 percent and the carbon footprint of its trucking fleet by several times this percentage. Additionally, progress is being made toward the firm's goal of zero waste going to landfills from its operations. A reduction of its waste by 81 percent in California suggests that this goal may be attainable. Competitor Target is also using its resources and capabilities for the purpose of forming a "green" core competence. "Environmental sustainability is integrated throughout our businesses-from the way we build our stores to the products on our shelves," the store says. Packaging its Archer Farms Balanced Potato Crisps in bags that are manufactured with 25 percent renewable plant-based plastic is one example of actions Target is taking to be environmentally sustainable.

The length of time a firm can expect to create value by using its core competencies is a function of how quickly competitors can successfully imitate a good, service, or process. Value-creating core competencies may last for a relatively long period of time only when all four of the criteria we discuss next are satisfied. Thus, either Walmart or Target would know that it has a core competence and possibly a competitive advantage in terms of green practices if the way the firm uses its resources to complete these practices satisfies the four criteria.

3. 10   Value Chain Analysis

Value chain analysis allows the firm to understand the parts of its operations that create value and those that do not. Understanding these issues is important because the firm earns above-average returns only when the value it creates is greater than the costs incurred to create that value.

The value chain is a template that firms use to analyze their cost position and to identify the multiple means that can be used to facilitate implementation of a chosen strategy. Today's competitive landscape demands that firms examine their value chains in a global rather than a domestic-only contest. In particular, activities associated with supply chains should be studied within a global context.

A value chain is a linked set of value-creating activities beginning with basic raw materials coming from suppliers, to a series of value-added activities involved in producing and marketing a product or service, and ending with distributors getting the final goods into the hands of the ultimate consumer. 价值链是指一系列具有内在联系的价值创造活动,开始于供应商提供的基本原材料,经过涉及产品或服务的生产和销售的一系列增值活动的转换,结束于将最终产品提供给最终消费者的经销商。

The value chains of most industries can be split into two segments: upstream and downstream halves. 大多数行业的价值链可以分为两个部分:上游部分和下游部分。

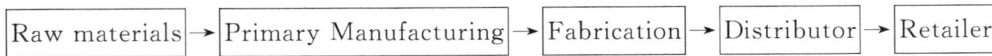

Raw materials → Primary Manufacturing → Fabrication → Distributor → Retailer

Figure 3.2　Typical Value Chain For a Manufactured Product

Table 3.6　Inter Factor Analysis Summary（IFAS）Table for Maytag

Internal factors	Weight	Rating	Weighted Score	Weighted Comments
Strengths				
• Quality Maytag Culture	0.15	5	0.75	Quality key to success
• Experienced top management	0.05	4	0.20	Know appliances
• Vertical integrations	0.10	4	0.40	Dedicated factories
• Employee relations	0.05	3	0.15	Goods, but deteriorating
• Hoover's international orientation	0.15	3	0.45	Hoover name in cleaners
Weaknesses				
• Process-oriented R&D	0.05	2	0.10	Slow on new products
• Distribution channels	0.05	2	0.10	Superstores replacing small dealers
• Financial position	0.15	2	0.30	High debt load
• Global positioning	0.20	2	0.40	Hoover weak outside the New Zealand, U.K., and Australia
• Manufacturing facilities	0.05	4	0.20	Investing now
Totals	1.00		3.05	

Each corporation has its internal value chain of activities. Porter proposes that a manufacturing firm's primary activities usually begin with inbound logistics(raw materials handling and warehousing), go through an operations process in which a product is manufactured, and continue to outbound logistics (warehousing and distribution), marketing and sales and finally to service(installation, repair, and sale of parts). 每家公司都有内部价值链活动。波特认为一家制造型企业的基础活动贯穿于产品制造的运行过程，经常开始于采购物流（原材料处理和仓储），继续于出境物流（仓储和分销），并最终服务于市场和销售（安装、维修和销售的部分）

Several support activities, such as procurement(purchasing), technology development (R&D), human resource management, and firm infrastructure(accounting, finance, and strategic planning), ensure that the primary value-chain activities operate effectively and efficiently. 几种辅助活动可保证公司价值链活动有效运行，比如采购、技术研发、人力资源管理和公司日常活动（会计，财务和战略计划）。

Corporate value-chain analysis involves the following steps：公司价值链分析包括以下步骤：

Examine each product line's value chain in terms of the various activities involved in producing that product or service. 根据生产产品或服务所涉及的各种活动检查每条产品线的价值链。

Examine the "linkages" within each product line's value chain. Linkage are the connections between the way one value activity (e. g. , marketing) is performed and the cost of performance of another activity(e. g. , quality control). 检验每条产品线的价值链的内在联系。联系是指执行某个价值活动的方式（例如营销）与其他活动（如质量控制）的绩效成本之间的关系。

Examine the potential synergies among the value chains of different product lines of business units. 检验不同产品线或业务单位价值链之间的潜在协同作用。

We show a model of the value chain in Figure 3.2. As depicted in the model, a firm's value chain is segmented into value chain activities and support functions. Value chain activities are activities of tasks the firm completes in order to produce products and then sell, distribute, and service those products in ways that create value for customers. Support functions include the activities or tasks the firm completes in order to support the work being done to produce, sell, distribute, and service the products the firm is producing. A firm can develop a capability and/or a core competence in any of the value chain activities and in any of the support functions. When it does so, it has established an ability to create value for customers. In fact, as shown in Figure 3.2, customers are the one firms seek to serve when using value chain analysis to identify their capabilities and core competencies. When using their unique core competencies to create unique value for customers that competitors cannot duplicate, firms have established one or more competitive advantages. This appears to be the case for P&G as it relies on the five core competencies described earlier in a Strategic Focus to produce unique, high-quality branded products that are sold to customers throughout the world.

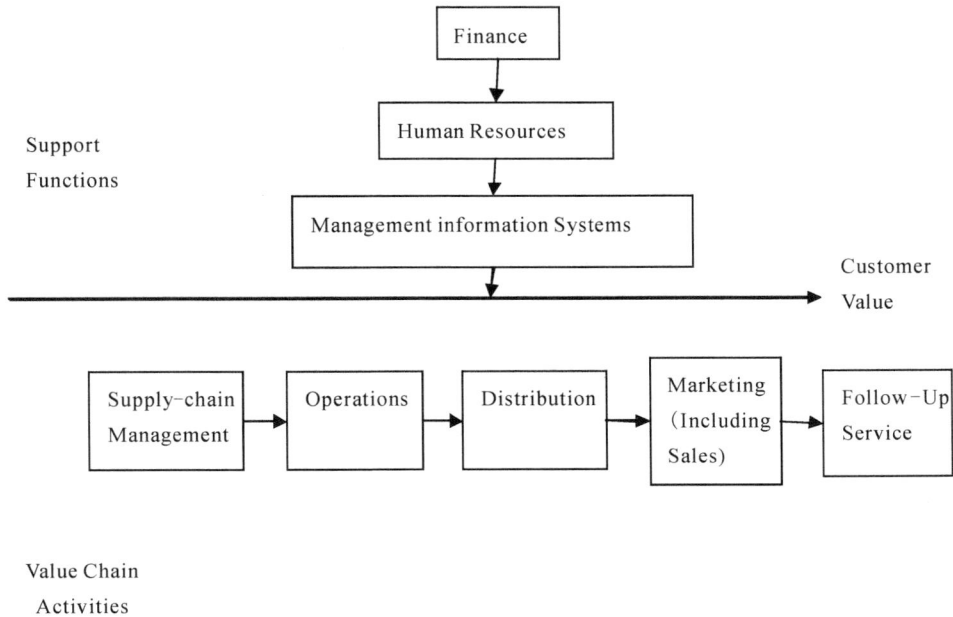

Support
Functions

Finance

Human Resources

Management information Systems

Customer
Value

| Supply-chain Management | Operations | Distribution | Marketing (Including Sales) | Follow-Up Service |

Value Chain
Activities

Figure 3. 3　A Model of the Value Chain

The activities associated with each part of the value chain are shown in Figure 3. 4 while the activities that are part of the tasks firms complete when dealing with support functions appear in Figure 3. 5. All items in both Figures should be evaluated relative to competitors' capabilities and core competencies. To become a core competence and a source of competitive advantage, a capability must allow the firm(1) to perform an activity in a manner that provides value superior to that provided by competitors, or(2) to perform a value creating activity that competitors cannot perform. Only under these conditions does a firm create value for customers and have opportunities to capture that that value.

Supply- Chain Management
Activities including sourcing procurement, conversion, and logistics management that are necessary for the firm to receive raw materials and convert them into final products.

Follow - up Service
Activities taken to increase a product's value for customers. Surveys to receive feedback about the customer's satisfaction, offering technical support after the sale, and fully complying with a product's warranty are examples of these activities.

Customer Value

Distribution
Activities related to getting the final product to the customer. Efficiently handling customer's orders, choosing the optimal delivery channel, and working with the finance support function to arrange for delivered goods are examples of these activities.

Marketing (Including Sales)
Activities taken for the purpose of segmenting target customers on the basis of their unique needs, satisfying customers' needs, retaining customers, and locating additional customers. Advertising campaigns, developing and managing product brands, determining appropriate pricing strategies, and training supporting a sales force are specific examples of these activities.

Operations
Activities necessary to efficiently change raw materials into finished products. Developing employees' work schedules, designing production processes and physical layout of the operations' facilities, determining production capacity needs, and selecting and maintaining production equipment sre examples of specific operations activities.

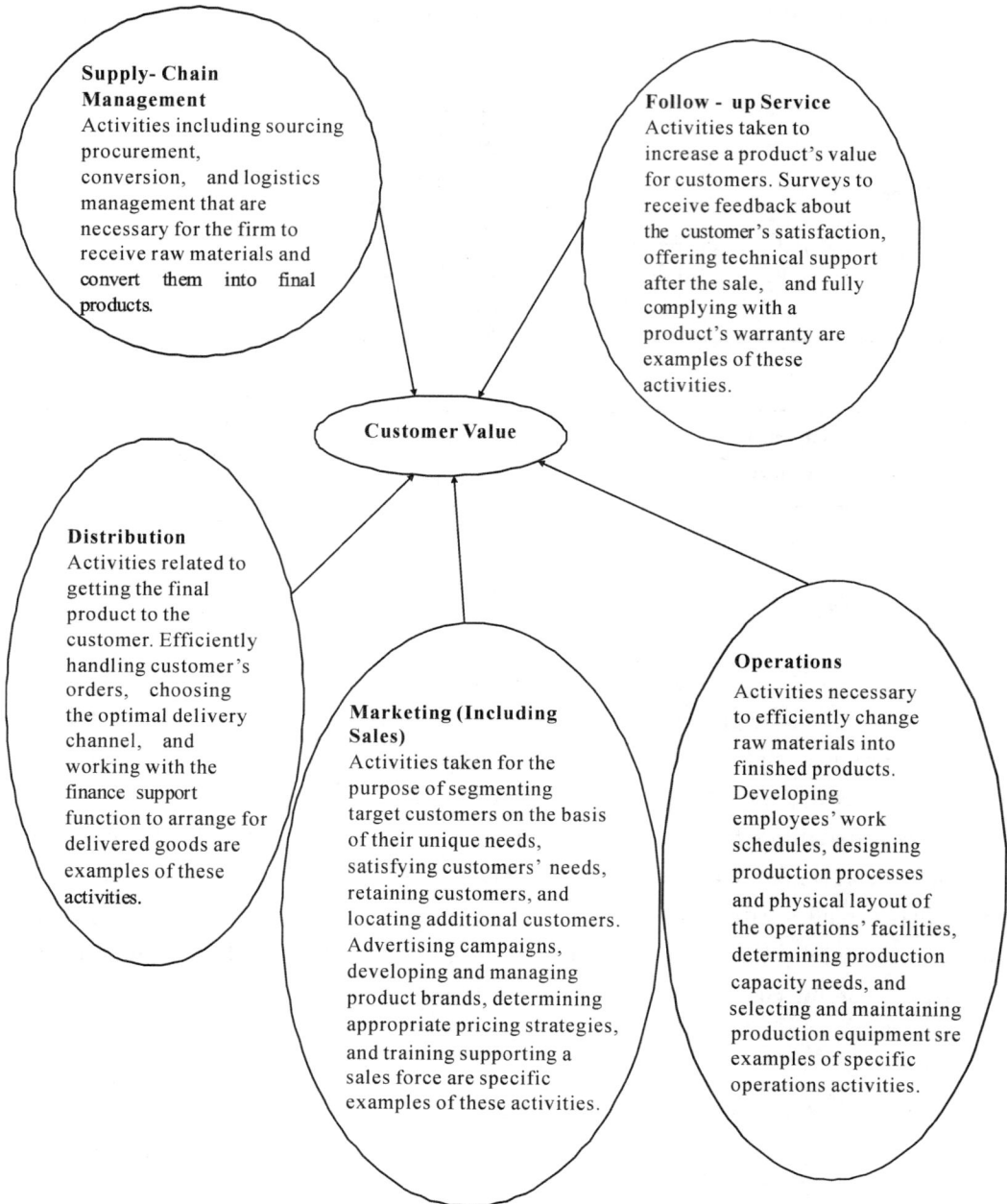

Figure 3. 4   Creating Value through Value Chain Activities

Human Resources
Activities associated with managing the firm's human resources in ways that create a capability and hopefully a core competence are specific examples of these activities.

Finance
Activities associated with effectively acquiring and managing financial reso urces. Securing adequate financial capital, investing in organizational functions in ways that will support the firm's efforts to produce and distribute its products in the short and long -term,   and managing relationship with those providing financial capital to the firm are specific examples of these activities.

Management Information Systems
Activities taken to obtain and manage information and knowledge throughout the firm. Identifying and utilizing sophisticated technologies, determining optimal ways to collect and distribute knowledge,   and linking relevant information and knowledge to organizational functions are activities associated with this support function.

Customer Value

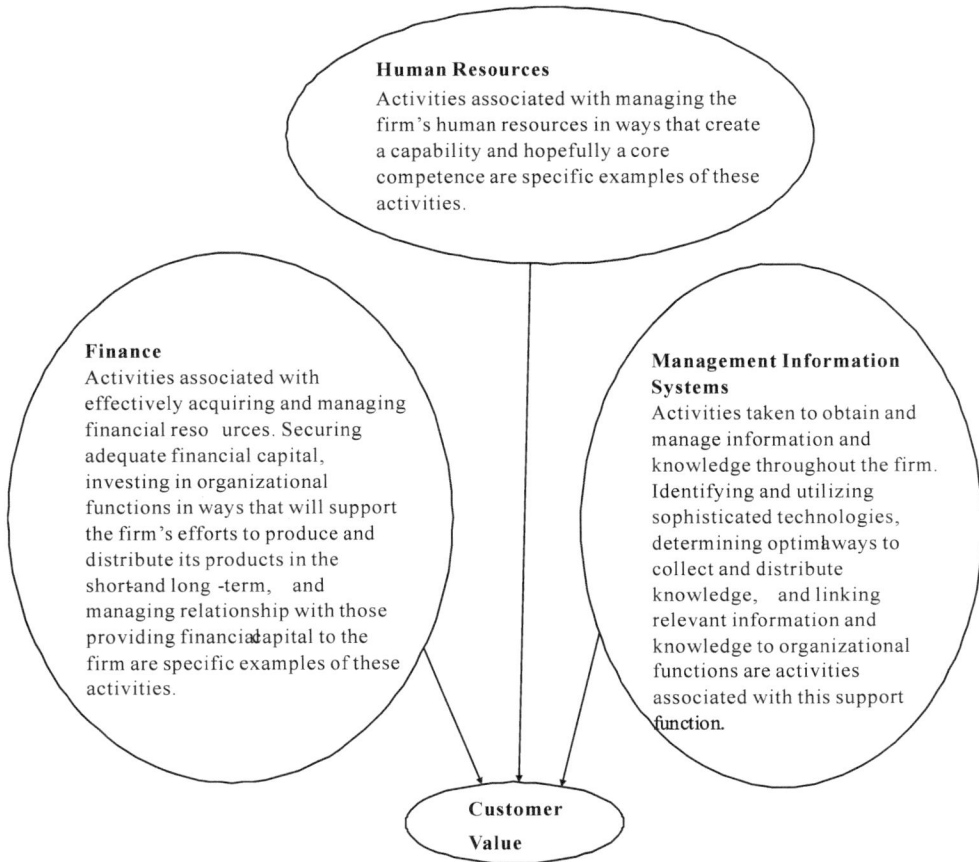

Figure 3. 5   Creating Value through Support Functions

Creating value for customers by completing activities that are part of the value chain often require building effective alliances with suppliers (and sometimes others to which the firm outsources activities, as discussed in the next section) and developing strong positive relationships with customers. When firms have such strong positive relationships with suppliers and customers, they ate said to have "Social capital". The relationships themselves have value because they produce knowledge transfer and access to resources that a firm may not hold internally. To build social capital whereby resources such as knowledge are transferred across organizations requires trust between the parties. The partners must trust each other in order to allow their resources to be used in such a way that both parties will benefit over time and neither party will take advantage of the other. Trust and social capital usually evolve over time repeated interactions but firms can also establish special means to jointly manage alliances that promote greater trust with the outcome of enhanced benefits for both partners.

Evaluating a firm's capability to execute its value chain activities and support functions is challenging. Earlier in the chapter, we noted that identifying and assessing the value of a firm's resources and capabilities requires judgment. Judgment is equally necessary when using value chain analysis, because no obviously correct model or rule is universally available to help in the process.

The analysis on enterprise value chain—(the teacher plans to use enlighten education way)

There are two kinds of activities which can make profits:

(1) Primary activates: Inbound Logistics, Operations, Outbound Logistics, Marketing and Sales, Service

(2) Support Activities: Firm Infrastructure, Human Resource Management, Technology Development, Procurement

2. Primary Activities

(1) Inbound Logistics

concept introduction—(the teacher plans to use enlighten education way)

Activities: such as material handling, warehousing and inventory control

case study—(the teacher plans to use case study way)

TOYOTA—JUST IN TIME relationship between manufacturer and supplier

(2) Operations

concept introduction—(the teacher plans to use enlighten education way)

Activities: such as machining, packaging, assembly

case study—(the teacher plans to use case study way)

Federal Express — effective material assembly

(3) Outbound Logistics

concept introduction—(the teacher plans to use enlighten education way)

Activities: such as collecting, storing, distributing the final product

case study—(the teacher plans to use case study and scenario education way)

7-11—the same trigger of many stores in one target city

(4) Marketing and Sales

concept introduction—(the teacher plans to use enlighten education way)

Activities: such as advertising, promotional campaigns, pricing

case study—(the teacher plans to use case study way)

CITROEN—The advertising campaign on new kind of car

(5) Service

concept introduction—(the teacher plans to use enlighten education way)

Activities: such as installation, repair, training

case study—(the teacher plans to use case study way)

Little Swan—the repair activity in the night

3. Support Activities

(1) Firm Infrastructure

concept introduction—(the teacher plans to use enlighten education way)

Activities: such as planning, finance, accounting

case study—(the teacher plans to use case study way)

P&G—patents strategy on shampoo product

(2) Human Resource Management

concept introduction—(the teacher plans to use enlighten education way)

Activities: such as recruiting, hiring and training

case study—(the teacher plans to use case study way)

HP—The new qualification need of a professional receptionist

(3) Technology Development

concept introduction—(the teacher plans to use enlighten education way)

Activities: such as process improvement, basic R&D, design

case study—(the teacher plans to use case study way)

TOYOTA—The development of Ford Company testifies the importance of R&D

(4) Procurement

concept introduction—(the teacher plans to use enlighten education way)

Activities: such as raw material and supplies, fixed assets equipment, buildings

case study—(the teacher plans to use case study way)

APP—cultivating the own tree resource

Enterprise competitive simulation—(the teacher plans to use commutative education way)

Enterprise competitive simulation design—the teacher classify the students as 10 groups, and choosing 5 kinds of industries (Computer, Microwave Oven, Cosmetics, Hotel Service, Medicine), let these 10 groups assemble fixed—quantity operating resources in the above 9 kinds of value chain activities, and then, making the combination of value chain operating resources several simulating unit of every groups, the teacher will give evaluation in operating situation of every groups.

Competitive Force of Buyers Buyers are a strong competitive force when: They are large and purchase a sizable percentage of industry's product. They buy in large quantities. They can integrate backward Industry's product is standardized. Their costs in switching to substitutes or other brands are low. They can purchase from several sellers Product purchased does not save buyer money.

Competitive Pressures: Collaboration Between Sellers and Buyers Partnerships are an increasingly important competitive element in business-to-business relationships. Collaboration may result in mutual benefits regarding. Just-in-time deliveries Order processing Electronic invoice payments. On-line sharing of sales at the cash register Competitive advantage potential may accrue to industry rivals who do the best job of managing seller-buyer partnerships.

Principle of Competitive Markets Buyers are a stronger competitive force the more they have leverage to bargain over: Price Quality Service Other terms and conditions of sale.

Strategic Implications of the Five Competitive Forces Competitive environment is unattractive from the standpoint of earning good profits when: Rivalry is strong Entry barriers are low and entry is likely Competition from substitutes is strong Suppliers and customers have considerable bargaining power.

Strategic Implications of the Five Competitive Forces Competitive environment is ideal from a profit-making standpoint when: Rivalry is moderate Entry barriers are high and no firm is likely to enter Good substitutes do not exist Suppliers and customers are in a weak bargaining position.

Coping With the Five Competitive Forces Objective is to craft a strategy. To insulate firm from competitive forces. To help make the "rules," placing added pressure on rivals Which allows firm to define the business model for the industry.

3.11   Analyzing Driving Forces

Identify those forces likely to exert greatest influence over next 1—3 years. Usually no more than 3—4 factors qualify as real drivers of change. Assess impact What difference will the forces make-favorable or unfavorable?

Common Types of Driving Forces Internet and e-commerce opportunities. Increasing globalization of industry Changes in long-term industry growth rate Changes in who buys the product and how they use it Product innovation Technological change/process innovation Marketing innovation.

Common Types of Driving Forces Entry or exit of major firms Diffusion of technical knowledge Changes in cost and efficiency Market shift from standardized to differentiated products (or vice versa) Regulatory policies/government legislation Changing societal concerns, attitudes, and lifestyles Changes in degree of uncertainty and risk.

Environmental Scanning Definition Monitoring and interpreting sweep of social, political，economic，ecological，and technological events to spot budding trends that could eventually impact industry Purpose. Raise consciousness of managers about potential developments that could Have important impact on industry conditions Pose new opportunities and threats.

Which Companies are in Strongest/Weakest Positions? One technique for revealing the different competitive positions of industry rivals is strategic group mapping. A strategic group consists of those rivals with similar competitive approaches in an industry.

Strategic Group Mapping Firms in same strategic group have two or more competitive characteristics in common Sell in same price/quality range. Cover same geographic are as be vertically integrated to same degree. Have comparable product line breadth Emphasize same types of distribution channels Offer buyers similar services Use identical technological approaches

Procedure for Constructing a Strategic Group Map STEP 1：Identify competitive characteristics that differentiate firms in an industry from one another. STEP 2：Plot firms on a two-variable map using pairs of these differentiating characteristics. STEP 3：Assign firms that fall in about the same strategy space to same strategic group. STEP 4：Draw circles around each group，making circles proportional to size of group's respective share of total industry sales.

Market position refers to the selection of specific areas for marketing concentration and can be expressed in terms of markets，product，and geographical locations. Through market research，corporations are able to practice markets segmentation—tailoring products for specific market niches. 市场定位是指选择营销活动集中的特定领域，可以表现在市场、产品和地理位置等方面。借助市场调查，企业可以进行市场细分——为特定利基市场提供适合的产品。

The product life cycle is a graph showing time plotted against the dollar sales of a product as it moves from introduction through growth and maturity to decline. 产品生命周期以图表形式展示了产品销售额（以美元计）随着时间推移而发生的变化，经历了引入、成长、成熟和衰退等阶段。

Strategic Group Maps Variables selected as axes should not be highly correlated Variables chosen as axes should expose big differences in how rivals compete. Variables do not have to be either quantitative or continuous. Drawing sizes of circles proportional to combined sales of firms in each strategic group allows map to reflect relative sizes of each strategic group. If more than two good competitive variables can be used，several maps can be drawn

Interpreting Strategic Group Maps Driving forces and competitive pressures often favor some strategic groups and hurt others Profit potential of different strategic groups varies due to strengths and weaknesses in each group's market position. The closer strategic groups are on map, the stronger the competitive rivalry among member firms tends to be

What Strategic Moves Are Rivals Likely to Make Next? A firm's own best strategic moves are affected by Current strategies of competitors Future actions of competitors Profiling key rivals involves gathering competitive intelligence about their Current strategies Most recent moves Resource strengths and weaknesses Announced plans.

A brand is a name given to a company's product which identifies that item in the mind of the consumer. 品牌是公司的产品名称在消费者心目中的标识。

A corporate brand is a type of brand in which the company's name serves as the brand. 企业品牌是指采用公司的名称作为品牌名称。

Competitor Analysis Successful strategists take great pains in scouting competitors to Understand their strategies Watch their actions Evaluate their vulnerability to driving forces and competitive pressures Size up their resource strengths and weaknesses and their capabilities Try to anticipate rivals' next moves

Predicting Moves of Rivals Predicting rivals' next moves involves. Analyzing their current competitive positions. Examining public pronouncements about what it will take to be successful in industry. Gathering information from grapevine about current activities and potential changes. Studying past actions and leadership. Determining who has flexibility to make major strategic changes and who is locked into pursuing same basic strategy.

The flexible manufacturing permits the low-volume output of custom-tailored products at relatively low unit costs through economies of scope. 柔性制造允许借助范围经济为客户定制产品，以较低的单位成本实现小批量产出。

Autonomous (self-managing) work teams in which a group of people work together without a supervisor to plan, coordinate, and evaluate their work. 自主工作团队又称为自我管理工作团队。在自主工作团队中，团队成员一起工作，没有负责计划、协调和评估工作的主管。

Over two-third of large U. S. firms are successfully using autonomous work teams. 超过2/3的美国大型公司成功使用了自主工作团队。

Companies have begun using cross-functional work teams instead of developing products in a series of steps-beginning with a request from sales, which leads to design, to engineering and to purchasing, and finally to manufacturing (often resulting in customer rejection of a costly product)—companies are tearing down the traditional walls separating departments so that people from each discipline can get involved in projects early on. 公司使用跨职能工作团队替代传统的产品开发方式，消除横亘在各部门之间、将各部门分离开来的传统界限，确保来自各个学科的人员在项目早期就参与其中。

Virtual teams are groups of geographically and / or organizationally dispersed coworkers that are assembled using a combination of telecommunications and information technologies to accomplish an organizational task. 虚拟团队是指利用电信和信息技术，将地理上和/或组织上分散的合作者组织起来，形成共同完成组织任务的团队。

Human diversity is the mix in the workplace of people from different races, cultures, and backgrounds. 人员多样性是指工作场所不同种族、文化和背景的人员的组合情况。

Supply chain management is the forming of networks for sourcing raw materials, manufacturing products or creating services, storing and distributing the goods, and delivering them to customers. 供应链管理是原材料采购、产品生产或服务创造、货物储存和分销，以及客户配送的网络形式。

Identifying Industry Key Success Factors Answers to three questions pinpoint KSFs On what basis do customers choose between competing brands of sellers? What resources and competitive capabilities does a seller need to have to be competitively successful? What does it take for sellers to achieve a sustainable competitive advantage? KSFs consist of the 3-5 really major determinants of financial and competitive success in an industry.

The concept of financial leverage (the ratio of total debt to total assets) helps describe the use of debt(versus equity)to finance the company's programs from outside. 财务杠杆概念帮助公司使用外部债务（相对于股权）进行融资的方案。

Capital budgeting is the analyzing and ranking of possible investments in fixed assets such as land, buildings, and equipment in terms of the additional outlays and additional receipts that will result from each investment. 资本预算是对固定资产，如土地、建筑物和设备，按照每项投资可能产生的额外支出和额外收入情况进行分析和排序。

A company's R&D intensity (its spending on R&D as a percentage of sales revenue) is a principal means of gaining market share in global competition. 一个公司的研发强度（即研发投入占销售收入的百分比）是在全球竞争中赢得市场份额的一种主要手段。

A company's R&D unit should be evaluated for technological competence, the proper management of technology, in both the development and the use of innovative technology. 评估一个公司研发部门的标准还包括技术竞争力、适当的技术管理，以及创新技术的发展和使用。

A company should also be proficient in technology transfer, the process of taking a new technology from the laboratory to the marketplace. 公司还应该擅长技术转让——新技术从实验室推向市场的过程。

The R&D mix involves basic R&D, product R&D, engineering or process R&D. 研发组合包括基础性研发、产品研发和工程或工艺研发。

The R&D mix is the balance of the three types of research. The mix should be appropriate to the strategy being considered and to each product's life cycle. 研发组合是三种研发类型研究的平衡。研发组合要适合所考虑的战略，以及每个产品的生命周期。

According to Richard Foster of McKinsey company, technological discontinuity is the displacement of one technology by another. 根据麦肯锡公司理查德·福斯特的研究，技术终结是一种技术替代另一种技术。

Even though the new technology may be more expensive to develop，it offers performance improvements in areas that are attractive to this small niche，but of no consequence to customers of the established competitors. 尽管新技术的开发可能更昂贵，但是新技术在很小的利基市场中所提供的绩效提升非常具有吸引力，对现有竞争对手的客户也没有什么影响。

Example：KSFs for Apparel Manufacturing Industry Fashion design—to create buyer appeal Low-cost manufacturing efficiency—to keep selling prices competitive.

Example：KSFs for Tin and Aluminum Can Industry Locating plants close to end-use customers—to keep costs of shipping empty cans low. Ability to market plant output within economical shipping distances

Things to Consider in Assessing Industry Attractiveness Industry's market size and growth potential Whether competitive conditions are conducive to rising/falling industry profitability. Will competitive forces become stronger or weaker. Whether industry will be favorably or unfavorably impacted by driving forces. Potential for entry/exit of major firms Stability/dependability of demand Severity of problems facing industry. Degree of risk and uncertainty in industry's future

Conducting an Industry and Competitive Situation Analysis Two things to keep in mind1. Evaluating industry and competitive conditions cannot be reduced to a formula-like exercise—thoughtful analysis is essential2. Sweeping industry and competitive analyses need to done every 1 to 3 years.

IFAS (Internal Factor Analysis Summary) table is one way to organize the internal factors into the generally accepted categories of strengths and weaknesses and to analyze how well a particular company's management is responding to these specific factors in light of the perceived importance of these factors to the company. 内部因素分析汇总表是将内部因素归类为普遍接受的优势和劣势，并分析特定公司的管理层是如何根据每个因素对于公司来说可感知的重要程度有所回应的。

Chapter 4 Evaluating Resources and Competitive Capabilities

4.1 A Model of Competitive Rivalry

Competitive rivalry evolves from the pattern of actions and responses as one firm's competitive actions have noticeable effects on competitors, eliciting competitive responses from them. This pattern suggests that firms are mutually interdependent, that they are affected by each other's actions and responses, and that marketplace success is a function of both individual strategies and the consequences of their use. Increasingly, too, executives recognize that competitive rivalry can have a major effect on the firm's financial performance. Research shows that intensified rivalry within an industry results in decreased average profitability for the competing firms. For example, Research in Motion (RIM) dominated the smartphone market with its Blackberry operating system platform until Apple's iPhone platform emerged. Likewise, the introduction of the Android platform by Google has cut into RIM's marked share and has thereby lowered the company's performance expectations.

Figure 4.1 presents a straightforward model of competitive rivalry at the firm level; this type of rivalry is usually dynamic and complex. The competitive actions and responses the firm takes are the foundation for successfully building and using its capabilities and core competencies to gain an advantageous market position. The model in Figure 4.1 presents the sequence of activities commonly involved in competition between a particular firm and each of its competitors' behavior and reduce the uncertainly associated with competitors' actions. Being able to predict competitors' actions and responses with competitors' actions. Being able to predict competitors' actions and responses has a positive effect on the firm's market position and its subsequent financial performance. The sum of all the individual rivalries modeled in Figure 4.1 that occur in a particular market reflect the competitive dynamics in that market.

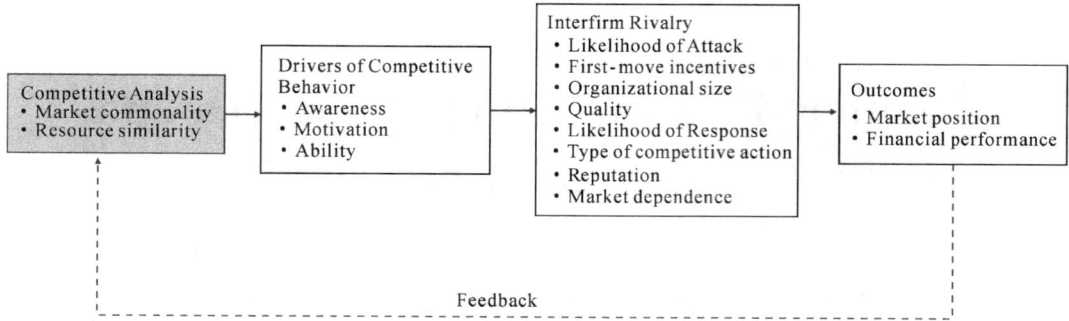

Figure 4.1   A Model of Competitive Rivalry

4.2   Competitor Analysis

As previously noted, a competitor analysis is the first step the firm takes to be able to predict the extent and nature of its rivalry with each competitor. The number of markets in which firms compete against each other and the similarity in their resources determine the extent to which the firms are competitors. Firms with high market commonality and highly similar resources are "clearly direct and mutually acknowledged competitors." The drivers of competitive behavior-as well as factors influencing the likelihood that a competitor will initiate competitive actions and will respond to its competitors' actions-influence the intensity of rivalry, even for direct competitors.

We discussed competitor analysis as a technique firms use to understand their competitive environment. Together, the general, industry, and competitive environments comprise the firm's external environment. We also described how competitor analysis is used to help the firm understand its competitors. This understanding results from studying competitors' future objectives, current strategies, assumptions, and capabilities.

Such competitive awareness is illustrated in the Strategic Focus on the competitors in the rivalry among the global automobile producers such as Toyota, Ford, General Motors, Honda, Chrysler, Nissan, and others. These analyses are highly important because they help managers to avoid "competitive blind spots," in which managers are unaware of specific competitors or their capabilities. If managers have competitive blind spots, they may be surprised by a competitor's actions, thereby allowing the competitor to increase its market share at the expense of the manager's firm. Competitor analyses are especially important when a firm enters a foreign market. Managers need to understand the local competition and foreign competitors currently operating in the market. Without such analyses, they are less likely to be successful.

4.3　Strategic and Tactical Actions

Firms use both strategic and tactical actions when forming their competitive actions and competitive response in the course of engaging in competitive rivalry. A competitive action is a strategic or tactical action the firm takes to build or defend its competitive advantages of improve its market position. A competitive response is a strategic or tactical action or a strategic response is a market-based move that involves a significant commitment of organizational resources and is difficult to implement and reverse. A tactical action or a tactical response is a market-based move that is taken to fine-tune a strategy; it involves fewer resources and is relatively easy to implement and reverse.

As noted in the Opening Case, Apple opened a service called "Game Center" once it found that users were using its iPhone, iPad, and iPod platforms for video games. With its update to its iOS (operating system) software, game producers began producing game applications to use the Apple system as its graphics became more advanced. This represents a strategic move by Apple. Game platform hardware and software producers such as Nintendo and Sony then created strategic responses to the Apple threat. For example, Sony, which produces the PlayStaion console, partnered with Sony Ericsson to make the Xperia PLAY phone, which uses "Play Station-certified games" and runs on Google's Android operating system.

Walmart prices aggressively as a mean of increasing revenues and gaining market share at the expense of competitors. However, pricing is a tactical strategy and Walmart's performance has lagged recently. In recent tactical moves, it has asserted that it will do a better job on competitive pricing. Although pricing aggressively is at the core of what Walmart is and how it competes, can the tactical action of aggressive pricing continue to lead to the competitive success the firm has historically enjoyed? Is Walmart achieving the type of balance between strategic and tactical competitive actions and competitive responses that is a foundation for all firm's success in marketplace competitions?

When engaging rivals in competition, firms must recognize the differences between strategic and tactical actions and responses and should develop an effective balance between the two types of competitive actions and responses. Airbus, Boeing's major competitor in commercial airliners, is aware that Boeing is strongly committed to taking actions it believers are necessary to successfully launch the 787 jetliner, because deciding to design, build and launch the 787 is a strategic action. In fact, many analysts believe that Boeing's development of the 787 airliner was a strategic response to Airbus's new A380 aircraft.

A corporation is a mechanism established to allow different parties to contribute capital, expertise, and labor for their mutual benefit. 公司是一种机制,其成立的目的是让

各方人士为了共同利益投入资金、技术和劳动力。

The boards of most publicly owned corporations are composed of both inside and outside directors. Inside directors are typically officers or executives employed by the corporation. Outside directors may be executives of other firms but are not employees of the board's corporation. 大多数上市公司的董事会由内部董事和外部董事构成。内部董事通常由公司聘用的主管或执行官担任。外部董事可能是其他公司的执行官,但不是本公司董事会聘用的雇员。

4.4　Competitive Dynamics

Whereas competitive rivalry concerns the ongoing actions and responses between a firm and its direct competitors for an advantageous market position, competitive dynamics concern the ongoing action actions and responses among all firms competing within a market for advantageous positions. Building and sustaining competitive advantages are at the core of competitive rivalry, in that advantages are the key to creating value for shareholders.

To explain competitive dynamics, we explore the effects of varying rates of competitive speed in different markets on the behavior of all competitors within a given market. Competitive behaviors as well as the reasons for taking them are similar within each market type, but differ across the three markets. Thus, competitive dynamic differ in slow-cycle, fast-cycle and standard-cycle markets. The sustainability of the firm's competitive advantages differs across the three market types. Research has also shown how firms go through life-cycle stages as markets evolve overtime within which a firm is competing. However, understanding what happens within each type of market is more pertinent in knowing how to respond to the competition.

Firms want to sustain their competitive advantages for as long as possible, although no advantage is permanently sustainable. The degree of sustainability is affected by how quickly by how quickly competitive advantages can be imitated and how costly it is to do so.

The managers of business organizations have four responsibilities: economic, legal, ethical, and discretionary. 商业组织的管理者有四个责任:经济责任、法律责任、道德责任和任意责任。

A corporation's task environment includes a large number of groups with interest in the activities of a business organization. These groups are called stakeholders because they are groups that affect or are affected by the achievement of the firm's objectives. 公司的任务环境中包括大量与组织的活动具有利益关系的群体。这些群体称为利益相关者,因为这些群体影响公司目标的实现或受到公司目标实现的影响。

4.5 Company Situation Analysis

How well is firm's present strategy working? What are the firm's resource strengths and weaknesses and its external opportunities and threats? Are firm's prices and costs competitive? How strong is firm's competitive position relative to rivals? What strategic issues does firm face?

How Well is the Present Strategy Working? Two steps involved Determine current strategy of company Examine key indicators of strategic and financial performance.

What is the Strategy? Identify competitive approach Low-cost leadership Differentiation Focus on a particular market niche Determine competitive scope Stages of industry's production/distribution chain Geographic coverage Customer base Identify functional strategies Examine recent strategic moves.

Key Indicators of How Well the Strategy is Working Trend in sales and market share. Acquiring and/or retaining customers Trend in profit margins Trend in net profits, ROI, and EVA Overall financial strength and credit ranking Efforts at continuous improvement activities Trend in stock price and stockholder value Image and reputation with customers Leadership role(s)—technology, quality, innovation, e-commerce, etc.

Question : What Are the Firm's Strengths, Weaknesses, Opportunities and Threats? SWOT represents the first letter in Strengths, Weaknesses, Opportunities, Threats for a company's strategy to be well-conceived, it must be matched to both Resource strengths and weaknesses Best market opportunities and external threats to its well-being SWOT.

Identifying Resource Strengths and Competitive Capabilities. A strength is something a firm does well or a characteristic that enhances its competitiveness Valuable competencies or know-how Valuable physical assets Valuable human assets Valuable organizational assets Valuable intangible assets Important competitive capabilities. An attribute that places a company in a position of market advantage Alliances or cooperative ventures with capable partners Resource strengths and competitive capabilities are competitive assets!

Mobilizing Company Resources to Produce Competitive Advantage Competitive Advantage Strategic Assets and Market Achievements Core and Distinctive Competencies Competitive Capabilities Company Resources.

Identifying Resource Weaknesses and Competitive Deficiencies. A weakness is something a firm lacks, does poorly, or a condition placing it at a disadvantage Resource weaknesses relate to Deficiencies in know-how or expertise or competencies. Lack of important physical, organizational, or intangible assets Missing capabilities in key areas Resource weaknesses and deficiencies are competitive liabilities!

The enterprise strategy explicitly articulates the firm's ethical relationship with its stakeholders. 企业战略阐明了企业与利益相关者之间的道德责任关系。

SWOT Analysis-What to Look For Potential Resource Strengths Potential Resource Weaknesses Potential Company Opportunities Potential External Threats? Powerful strategy? Strong financial condition? Strong brand name image/reputation? Widely recognized market leader? Proprietary technology? Cost advantages? Strong advertising? Product innovation skills? Good customer service? Better product quality? Alliances or JVs? No clear strategic direction? Obsolete facilities? Weak balance sheet; excess debt? Higher overall costs than rivals? Missing some key skills/competencies? Subpar profits? Internal operating problems...? Falling behind in R&D? Too narrow product line? Weak marketing skills? Serving additional customer groups? Expanding to new geographic areas? Expanding product line? Transferring skills to new products? Vertical integration? Take market share from rivals? Acquisition of rivals? Alliances or JVs to expand coverage? Openings to exploit new technologies? Openings to extend brand name/image? Entry of potent new competitors? Loss of sales to substitutes? Slowing market growth? Adverse shifts in exchange rates &trade policies? Costly new regulations? Vulnerability to business cycle? Growing leverage of customers or suppliers? Reduced buyer needs for product? Demographic changes.

TOWS (SWOT backwards) Matrix illustrates how the external opportunities and threats facing a particular corporation can be matched with that company's internal strengths and weaknesses to result in four sets of possible strategic alternatives. TOWS 矩阵(SWOT 分析的反向形式)描述了特定公司所面临的外部的机会和威胁与公司内部的优势和劣势之间的匹配关系,可以产生四种可能的备选战略集。

Competencies vs. Core Competencies vs. Distinctive Competencies. A company competence is the product of organizational learning and experience and represents real proficiency in performing an internal activity. A core competence is a well-performed internal activity that is central(not peripheral or incidental)to a company's competitiveness and profitability. A distinctive competence is a competitively valuable activity that a company performs better than its rivals.

Company Competencies and Capabilities Stem from skills, expertise, and experience usually representing an Accumulation of learning over time and Gradual buildup of real proficiency in performing an activity. Involve deliberate efforts to develop the ability to do something, often entailing Selection of people with requisite knowledge and expertise Upgrading or expanding individual abilities. Molding work products of individuals into a cooperative effort to create organizational ability. A conscious effort to create intellectual capital.

Core Competencies：A Valuable Company Resource A competence becomes a core competence when the well-performed activity is central to the company's competitiveness and profitability. Often，a core competence results from collaboration among different parts of an organization. Typically，core competencies reside in a company's people，not in assets on the balance sheet. A core competence gives a company a potentially valuable competitive capability and represents a definite competitive asset.

The attractiveness of a particular strategic alternative is partially a function of the amount of risk it entails. Risk is composed not only of the probability that the strategy will be effective，but also of the amount of assets will be unavailable for other uses. 特定备选战略方案是否具有吸引力部分在于方案所涉及的大量风险。风险不仅包括战略结果最终是否有效，还包括公司必须分配给该战略的资产总额，以及这些资产不能为其他战略使用的时间长度。

The attractiveness of a strategic alternative is affected by its perceived compatibility with the key stakeholders in a corporation's task environment. 备选战略方案的吸引力受到公司与任务环境中关键利益相关者可感知的兼容性的影响。

Types of Core Competencies Expertise in building networks and systems to enable e-commerce Speeding new/next-generation products to market Better after-sale service capability Skills in manufacturing a high quality product Innovativeness in developing popular product features. Speed/agility in responding to new market trends System to fill customer orders accurately and swiftly Expertise in integrating multiple technologies.

Distinctive Competence—A Competitively Superior Resource ♯ 1A distinctive competence is a competitively significant activity that a company performs better than its competitors. A distinctive competence Represents a competitively valuable capability rivals do not have Presents attractive potential for being a cornerstone of strategy. Can provide a competitive edge in the marketplace—because it represents a competitively superior resource strength.

Strategic Management Principle A distinctive competence empowers a company to build competitive advantage！

Examples：Distinctive Competencies Sharp Corporation Expertise in flat-panel display technology Toyota，Honda，Nissan Low-cost，high-quality manufacturing capability and short design-to-market cycles Intel Ability to design and manufacture ever more powerful microprocessors for PCs Motorola Defect-free manufacture（six-sigma quality）of cell phones.

Determining the Competitive Value of a Company Resource To qualify as the basis for sustainable competitive advantage，a "resource" must pass 4 tests1. Is the resource hard to copy? Does the resource have staying power—is it durable? Is the resource really competitively superior? 4. Can the resource be trumped by the different capabilities of rivals?

Strategic Management Principle Successful strategists seek to capitalize on and

71

leverage a company's resource strengths—its expertise, core competencies, and strongest competitive capabilities—by molding the strategy around the resource strengths!

Identifying a Company's Market Opportunities Opportunities most relevant to a company are those offering Best prospects for profitable long-term growth Potential for competitive advantage Good match with its financial and organizational resource capabilities.

Strategic Management Principle. A company is well-advised to pass on a particular market opportunity unless it has or can build the resource capabilities to capture it!

Identifying External Threats Emergence of cheaper/better technologies Introduction of better products by rivals Intensifying competitive pressures Onerous regulations Rise in interest rates Potential of a hostile take over Unfavorable demographic shifts Adverse shifts in foreign exchange rates Political upheaval in a country.

If a strategy is incompatible with the corporate culture, it probably will not succeed. 如果一家公司的战略与企业文化不相适应,那么这个战略可能不会成功。

In considering a strategic alternative, strategy makers must assess its compatibility with the corporate culture. 考虑到战略的备选性,战略制定者一定会评估战略与企业文化的适应程度。

Strategic Management Principle Successful strategists aim at capturing a company's best growth opportunities and creating defenses against external threats to its competitive position and future performance!

Role of SWOT Analysis in Crafting a Better Strategy Developing a clear understanding of a company's Resource strengths Resource weaknesses Best opportunities External threats Drawing conclusions about how Company's strategy can be matched to both its resource capabilities and market opportunities Urgent it is for company to correct resource weaknesses and guard against external threats.

Are the Company's Prices and Costs Competitive? Assessing whether a firm's costs are competitive with those of rivals is a crucial part of company analysis Key analytic tools Strategic cost analysis Value chain analysis Benchmarking.

Why Rival Companies Have Different Costs Companies do not have the same costs because of differences in Prices paid for raw materials, component parts, energy, and other supplier resources Basic technology and age of plant & equipment Economies of scale and experience curve effects Wage rates and productivity levels Marketing, promotion, and administration costs Inbound and outbound shipping costs Forward channel distribution costs.

Principle of Competitive Markets The higher a company's costs are above those of close rivals, the more competitively vulnerable it becomes!

What is Strategic Cost Analysis? Focuses on a firm's costs relative to its rivals

Compares a firm's costs activity by activity against costs of key rivals From raw materials purchase to Price paid by ultimate customer Pinpoints which internal activities are a source of cost advantage or disadvantage.

The Concept of a Company Value Chain. A company consists of all the activities and function sit performs in trying to deliver value to its customers. A company's value chain shows the linked set of activities, functions, and business processes that it performs in the course of designing, producing, marketing, delivering, and supporting its product/service and thereby creating value for its customers. A company's value chain consists of two types of activities Primary activities (where most of the value for customers is created) Support activities that are undertaken to aid the individuals ands groups engaged in doing the primary activities.

4. 6   Typical Company Value Chain

The Value Chain System for an Entire Industry. Assessing a company's cost competitiveness involves comparing costs all along the industry's value chain Suppliers' value chains are relevant because Costs, quality, and performance of inputs provided by suppliers influence a firm's own costs and product performance Forward channel allies' value chains are relevant because. Forward channel allies' costs and margins are part of price paid by ultimate end-user Activities performed affect end-user satisfaction.

Example: Key Value Chain Activities SOFT DRINK INDUSTRY Processing of basic ingredients Syrup manufacture Bottling and can filling Wholesale distribution Retailing Kroger.

Activity-Based Costing: A Key Tool in Strategic Cost Analysis Determining whether a company's costs are inline with those of rivals requires measuring how a company's costs compare with those of rivals activity-by-activity—from one end of the value chain to the other Requires having accounting data that measures the cost of each value chain activity. Activity-based accounting systems provide the data for determining the costs for each relevant value chain activity.

Benchmarking Costs of Key Value Chain Activities Focuses on cross-company comparisons of how certain activities are performed and the costs associated with these activities. Purchase of materials Payment of suppliers Management of inventories. Training of employees Processing of payrolls Getting new products to market Performance of quality control Filling and shipping of customer orders.

Even the most attractive might not be selected if it is contrary to the needs and desires of important managers. 即使这个战略是最有吸引力,但如果它与重要管理者的需求和愿望相违背,那么它也可能不会被选择。

Strategic choice is the evaluation of alternative strategies and the selection of the best alternative. 战略选择是评价备选战略方案并选择最佳战略方案的过程。

Objectives of Benchmarking Determine whether a company is performing particular value chain activities efficiently by studying the practices and procedures used by other companies. Understand the best practices in performing an activity—learn what is the "best" way to do a particular activity from those who have demonstrated they are "best-in-industry" or "best-in-world" Assess if company's costs of performing particular value chain activities are in line with competitors. Learn how other firms achieve lower costs. Take action to improve company's cost competitiveness.

Ethical Standards in Benchmarking：Do's and Don'ts Avoid talk about pricing or competitively sensitive costs. Don't ask rivals for sensitive data. Don't share proprietary data without clearance. Have impartial third party assemble and present competitively sensitive cost data with no names attached. Don't disparage a rival's business to outsiders based on data obtained.

What Determines Whether a Company is Cost Competitive? A company's cost competitiveness depends on how well it manages its value chain relative to how well competitors manage their value chains. When a company's costs are "out-of-line", the "high-cost" activities can exist in any of three areas in the industry value chain1. Suppliers' activities2. The company's own internal activities3. Forward channel activities Activities, Costs, & Margins of Forward Channel Allies & Strategic Partners Internally Performed Activities, Costs, & Margins. Activities, Costs, & Margins of Suppliers Buyer/ User Value Chains.

Correcting Supplier-Related Cost Disadvantages：Options Negotiate more favorable prices with suppliers Work with suppliers to help them achieve lower costs. Use lower-priced substitute inputs Collaborate closely with suppliers to identify mutual cost-saving opportunities Integrate backwards Make up difference by initiating cost savings in other areas of value chain.

Correcting Forward Channel Cost Disadvantages：Options Push for more favorable terms with distributors and other forward channel allies Work closely with forward channel allies and customers to identify win-win opportunities to reduce costs Change to a more economical distribution strategy Make up difference by initiating cost savings earlier in value chain.

Correcting Internal Cost Disadvantages：Options Reengineer how the high-cost activities or business processes are performed Eliminate some cost-producing activities altogether by revamping value chain system. Relocate high-cost activities to lower-cost geographic areas. See if high-cost activities can be performed cheaper by outside vendors/ suppliers Invest in cost-saving technology Simplify product design. Make up difference by achieving savings in backward or forward portions of value chain.

From Value Chain Analysis to Competitive Advantage A company can create competitive advantage by managing its value chain to Integrate knowledge and skills of employees in competitively valuable ways. Leverage economies of learning/experience Coordinate related activities in ways that build valuable capabilities. Build dominating expertise in a value chain activity critical to customer satisfaction or market success.

From Value Chain Analysis to Competitive Advantage Strategy-Making Lesson of Value Chain Analysis Sustainable competitive advantage can be created by1. Managing value chain activities better than rivals and/or. Developing distinctive value chain capabilities to serve customers!

Chapter 5　Business-Level Strategies

5.1　Customers：Their Relationship with Business-Level Strategies

Strategic competitiveness results only when the firm satisfies a group of customers by using its competitive advantages as the basis for competing in individual product markets. A key reason firms must satisfy customers with their business-level strategy is that returns earned from relationships with customers are the lifeblood of all organizations.

The most successful companies try to find new ways to satisfy current customers and/or to meet the needs of new customers. Being able to do this can be even more difficult when firms and consumers face challenging economic conditions. During such times, firms may decide to reduce their workforce to control costs. This can lead to problems, however, when having fewer employees makes it more difficult for companies to meet individual customers' needs and expectations. In these instances, some suggest that firms should follow several courses of action, including paying extra attention to their best customers and developing a flexible workforce by cross-training employees so they can undertake a variety of responsibilities on their jobs. Amazon.com, insure USAA, and Lexus have been identified as "customer service champions" because they devote extra care and attention to customer service especially during challenging economic times.

Corporate strategy deals with three key issues facing the corporation as a whole that directional strategy, portfolio strategy and parenting strategy. 公司战略面对整个公司，把整个公司作为一个整体处理三个关键问题，即定向战略的问题、投资组合战略的问题及母合战略的问题。

Corporate strategy is therefore concerned with the direction of the firm and the management of its product lines and business units. This is true whether the firm is a small, one-product company or a large, multinational corporation. 公司战略主要关注企业的发展方向并管理公司的产品线和业务部门。无论这家公司是一家小型的、单一产品的企业还是大型的、多样化生产的企业，公司战略对它们而言都是真实存在的。

5.2　The Purpose of Business-Level Strategies

The purpose of a business-level strategy is to create differences between the firm's position and those of its competitors. To position itself differently from competitors, a firm must decide whether it intends to perform activities differently or to perform different activities. Strategy defines the path which provides the direction of actions to be taken by leaders of the organization. In fact, "choosing to perform activities differently or to perform different activities than rivals" is the essence of business-level strategy. Thus, the firm's business-level strategy is a deliberate choice about how it will perform the value chain's primary and support activities to create unique value. Indeed, in the current complex competitive landscape, successful use of a business-level strategy results from the firm learning how to integrate the activities it performs in ways that create superior value for customers.

Firms develop an activity map to show how they integrate the activities they perform. We show the Southwest Airlines activity map. The manner in which Southwest has integrated its activities is the foundation for the successful use of its primary cost leadership strategy (this strategy is discussed later in the chapter) but also includes differentiation through the unique services provided to customers. The tight integration among Southwest's activities is a key source of the firm's ability to at least historically operate more profitably than its competitors.

A corporation's directional strategy is composed of three general orientations toward growth: growth strategies, stability strategies, retrenchment strategies. 一个公司的定位战略是由增长战略、稳定战略或收缩战略这三个面对增长的总体方向组成的。

Southwest Airlines has conFigured the activities it performs into six strategic themes-limited passenger service; frequent, reliable departures; lean, highly productive ground and gate crews; high aircraft utilization; very low ticket prices; and short-haul. Point-to-point routes between mid-sized cities and secondary airports. Individual clusters of tightly linked activities make it possible for the outcome of a strategic theme to be achieved. For example, no meals, no seat assignments, and no baggage transfers form a cluster of individual activities that support the strategic theme of limited passenger.

Southwest's tightly integrated activities make it difficult for competitors to imitate the firm's cost leadership strategy. The firm's unique culture and customer service, both of which are sources of competitive advantages, are features that rivals have been unable to imitate, although some have tried and largely failed (e. g., U. S. Airways' Metrojet subsidiary, United Shuttle, Delta's Song, and Continental Airlines' Continental Lite). Hindsight shows that these competitors offered low price to customers, but weren't able to operate at costs close to those of Southwest or to provide customers with any notable sources of differentiation, such as a unique experience while in the air. The key to Southwest's success has been its ability to continuously reduce its costs while providing customers with acceptable levels of differentiation such as an engaging culture. Firms using the cost leadership strategy must understand that in terms of sources of differentiation that accompany the cost leader's product，the customer defines acceptable. Fit among activities is a key to the sustainability of competitive advantage for all firms， including Southwest Airlines. Strategic fit among the many activities is critical for competitive advantage. It is more difficult for a competitor to match a configuration of integrated activities than to imitate a particular activity such as sales promotion，or a process technology.

Strategy formulation is often referred to as strategic planning or long-range planning and is concerned with developing a corporation's mission，objectives，strategies，and policies. 战略制定通常是指战略规划或长期规划，主要关注开发公司的使命、目标、战略和政策。

The SFAS (Strategic Factors Analysis Summary) Matrix summarize a corporation's strategic factors by combining the external factors from the EFAS Table with the internal factors form the IFAS Table. 战略因素分析汇总矩阵通过将外部因素分析汇总表中的外部因素与内部因素分析汇总表中的内部因素相结合，总结一家公司的战略因素。

Table 5.1　Strategic Factors Analysis Summary (SFAS) Matrix

Strategic Factors (Select the most important opportunities/threats from the EFAS，Table 3-3，and the most important strengths and weaknesses from the IFAS，Table 4-2)	Weight	Rating	Weight Score	Duration			comments
				short	intermediate	long	
• Quality Maytag Culture(S)	0.10	5	0.50			×	Quality key to success
• Hoover's international orientation(S)	0.10	3	0.30		×		Name recognition
• Financial position(W)	0.10	2	0.20		×		High debt
• Global positioning(W)	0.15	2	0.30			×	Only in New Zealand, U.K.，and Australia
• Economic integration of European Union(O)	0.10	4	0.40			×	Acquisition of Hoover
• Demographics favor quality(O)	0.10	5	0.50		×		Maytag quality
• Trend to superstores(O+T)	0.10	2	0.20	×			Weak in this channel
• Whirlpool and Electrolux(T)	0.15	3	0.45	×			Dominate industry
• Japanese appliance companies (T)	0.10	2	0.20			×	Asian presence
Total	1.00		3.05				

Corporate scenarios are pro forma balance sheets and income statements that forecast the effect that each alternative strategy and its various programs will likely have on division and corporate return on investment. 公司情境是使用预计资产负债表和利润表,预测每个备选战略及其方案可能对分部和公司的投资收益产生的影响。

Develop common-financial statements for the company's or business unit's previous years. 为公司或业务单位前些年的财务情况制定共同比财务报表。

Construct detailed pro forma financial statements for each strategic alternative. 为每一个备选战略构建详细的预测财务报表。

One approach is to appoint someone as devil's advocate, a person or group assigned to identify potential pitfalls and problems with a proposed alternative. Another approach, called dialectical inquiry, requires that two proposals using different assumptions be generated for each alternative strategy under consideration. 一种方法是任命某人为魔鬼代言人,指出其他人或群体提出的备选方案的潜在陷阱和问题。另一种方法称为辩证质询,要求两个提案采用不同的假设。

When crafted correctly, an effective policy accomplishes three things: 一项有效的政策被正确的执行和完成需要注意三件事:

It forces trade-offs between competing resource demands. 推动竞争性资源需求之间的权衡。

It tests the strategic soundness of a particular action. 测试特定战略行动的稳健性。

It sets clear boundaries within which employees must operate while granting them freedom to experiment within those constraints. 设置清晰的员工工作界限,同时赋予员工在允许范围内进行尝试的自由。

5.3　Types of Business-Level Strategies

Firms choose from among five business-level strategies to establish and defend their desired strategic position against competitor: cost leadership, differentiation, focused cost leadership. Focused differentiation, and integrated cost leadership/ differentiation. Each business-level strategy helps the firm to establish and exploit a particular competitive advantage within a particular competitive scope. How firms integrate the activities they perform within each different business-level strategy demonstrates how they differ from one another. For example, firms have different activity maps, and thus, a Southwest Airlines activity map differs from those of competitors JetBlue, Continental, American Airlines, and so forth. Superior integration of activities increases the likelihood of being able to gain an advantage over competitors and to earn above-average returns.

Basis for customer value

	Lowest Cost	Distinctiveness
Broad Market	Cost Leadership	Differentiation
Target Market	Integrated Cost Leadership/ Differentiation	
Narrow Market Segment(s)	Focused Cost Leadership	Focused Differentiation

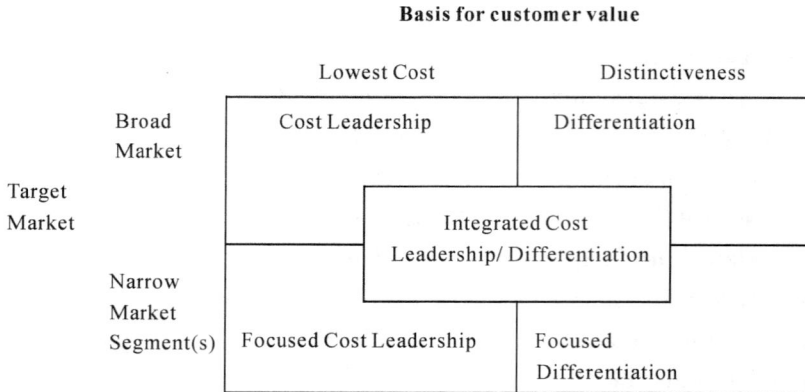

Figure 5. 1   Five Business-Level Strategies

Business strategies focuses on improving the competitive position of a company's or business unit's products or services within the specific industry or market segment that the company of business unit serves. 业务战略关注改善公司或业务单位涉足的特定行业或细分市场中产品或服务的竞争地位。

Competitive strategy creates a defendable position in an industry so that a firm can outperform competitors. 竞争战略巩固了公司在行业中的防御地位，从而超越竞争对手。

Michael Porter proposes two "generic" competitive strategies for outperforming other corporations in a particular industry: lower cost and differentiation. These strategies are called generic because they can be pursued by any type or size of business firm, even by not-for-profit organizations. 迈克尔·波特认为有两种通用的竞争战略能在一个行业中有助于企业跑赢其他公司：低成本战略和差异化战略。它们之所以被称为通用战略是因为它们能被任何类型或规模的商业公司或非营利组织所利用。

By far the most widely pursued corporate strategies of business firms are those designed to achieve growth in sales, assets, profits, or some combination of these. there are two basic corporate growth strategies: concentration within one product line or industry and diversification into other products or industries. 到目前为止，最广泛追求的企业战略的是那些旨在实现销售，资产，利润，或一些组合增长的一些战略。有两种基本的企业成长战略，一种是集中于一个产品或一个产业的增长战略，另一种是在其他产品或行业多元化发展的增长战略。

5. 4   Cost Leadership Strategies

The cost leadership strategy is an integrated set of actions taken to produce goods or services with features that are acceptable to customers at the lowest cost, relative to those of competitors. Firms using the cost leadership strategy commonly sell standardized goods or services (but with competitive levels of differentiation) to the industry's most typical customers. Process innovations, which are newly designed production and distribution methods and techniques that allow the firm to operate more efficiently, are critical to successful use of the cost leadership strategy.

Lower cost strategy is the ability of a company or a business unit to design, produce, and market a comparable product more efficiently than its competitors. 低成本战略是一个公司或业务单位在设计、生产和销售产品中表现出来的比其他竞争者更高效的能力。

As noted, cost leaders' goods and services must have competitive levels of differentiation that create value for customers. For example, in recent years Kia Motors has emphasized the design of its cars in the U. S. market as a source of differentiation while implementing a cost leadership strategy. Called "cheap chic", some analysts had a positive view of this decision, saying that "When they're done, Kia's cars will still be low-end (in price), but they won't necessarily look like it". It is important for firms using the cost leadership strategy to not only concentrate on reducing costs because it could result in the firm efficiently producing products that no customer wants to purchase. In fact, such extremes could limit the potential for important process innovations and lead to employment of lower-skilled workers, poor conditions on the production line, accidents, and a poor quality of work life for employees.

The firm using the cost leadership strategy targets a broad customer segment or group. Cost leaders concentrate on finding ways to lower their costs relative to competitors by constantly rethinking how to complete their primary and support activities to reduce costs still further while maintaining competitive levels of differentiation.

For example, cost leader Greyhound lines Inc. continuously seeks way to reduce the costs it incurs to provide bus service while offering customers have while they pay low prices for their service package. Interestingly, a number of customers have while they pay low prices trying to enhance the value of the experience customers have while they pay low prices for their service package. Interestingly, a number of customers now "insist on certain amenities that they receive on planes and trains-such as Internet access and comfortable seats, not to mention cleanliness". To maintain competitive levels of differentiation while using the cost leadership strategy, Greyhound recently starting using over 100 "motor coaches" that have leather seats, additional legroom, Wi-Fi access, and power outlets in every row.

Greyhound enjoys economies of scale by serving more than 25million passengers annually with about 2300 destinations in the United States and almost 13000 daily departures. These scale economies allow the firm to keep its costs low while offering some of the differentiated services today's customers seek from the company. Demonstrating the firm's commitment to the physical environment segment of the general environment is the fact that "one Greyhound bus takes an average of 35 cars off the road".

Cost leadership is a low-cost competitive strategy that aims at the broad mass market and requires "aggressive construction of efficient-scale facilities, vigorous pursuit of cost reductions from experience, tight cost and overhead control, avoidance sales force, advertising, and so on". 成本领先是一种瞄准广阔的大众市场采取低成本方式竞争的战略,需要"积极建设高效的规模化设施,借助经验、成本和额外费用的严格控制,大力追求成本降低,避免客户账户边际化,在研发、服务、销售队伍、广告等领域实现成本最小化,等等"。

As primary activities, inbound logistics (e. g., materials handling, warehousing, and inventory control) and outbound logistics (e. g., collecting, storing, and distributing products to customers) often account for significant portions of the total cost to produce some goods and services. Research suggests that having a competitive advantage in logistics creates more with a cost leadership than with a differentiation strategy. Thus, cost leaders seeking competitively valuable ways to reduce costs may want to concentrate on the primary activities of inbound logistics and outbound logistics. In so doing many firms choose to outsource their manufacturing operations to low-cost firms with low-wage employees (e. g., China). However, care must be taken because outsourcing also makes the firm more dependent on firms over which they have little control. At best, it creates interdependencies between the outsourcing firm and the suppliers. If dependencies become too great, it gives the supplier more power with which the supplier may increase price of the goods and services provided. Such actions could harm the firm's ability to maintain a low-cost competitive advantage.

Cost leader also carefully examine all support activities to find additional potential cost reductions. Developing new systems for finding the optimal combination of low cost and acceptable levels of differentiation in the raw materials required to produce the firm's goods or services is an example of how the procurement support activity can facilitate successful use of the cost leadership strategy.

Big Lost Inc. use the cost leadership strategy. With its vision of being "The World's Best Bargain Place, " Big Lost is the large closeout retailer in the United States with annual sales of over5 billion from more than 1500 stores with approximately 13, 000 employees. For Big Lost, closeout goods are brand-name products sold by other retailers provided for sale at substantially lower prices.

Firms use value-chain analysis to identify the parts of the company's operations that create value and those that do not. The primary and support activities that allow a firm to create value through the cost leadership strategy. Companies unable to link the activities shown in this Figure through the activity map they form typically lack the core competencies needed to successfully use the cost leadership strategy.

Effective use of the cost leadership strategy allows a firm to earn above-average returns in spite of the presence of strong competitive force. The next sections explain how firms implement a cost leadership strategy.

5.5　Differentiation Strategies

The differentiation strategy is an integrated set of action taken to produce goods or services (at an acceptable cost) that customers perceive as being different in ways that are important to them. While cost leaders serve a typical customer in an industry, differentiators target customers for whom value is created by the manner in which the firm's products differ from those produced and marketed by competitors. Product innovation, which is "the result of bringing to life a new way to solve the customer's problem-through a new product or service development-that benefits both the customer and the sponsoring company is critical to successful use of the differentiation strategy".

Differentiation strategy is the ability to provide unique and superior (value) to the buyer in terms of product quality, special features, or after-sale service. 差异化战略是能长期为买方在产品质量、特殊功能或售后服务中提供独特的超价值的这样一种能力。

Porter further proposes that a firm's competitive advantage in an industry is determined by its competitive scope, that is, the breadth of the target market of the company or business unit. 波特预见一个公司在产业中的竞争优势取决于它的竞争范围，也就是这个公司或业务单位的目标市场的广度。

Firms must be able to produce differentiated products at competitive costs to reduce upward pressure on the price that customers pay. When a product's differentiated features are produced at noncompetitive costs, the price for the product may exceed what the firm's target customers are willing to pay. If the firm has a thorough understanding of what its target customers value, the relative importance they attach to the satisfaction of differentiation strategy can be effective in helping it earn above-average returns. Of course, to achieve these returns, the firm must apply its knowledge capital to provide customers with a different product that provides them with superior value.

Differentiation is aimed at the broad mass market and involves the creation of a product or service that is perceived throughout its industry as unique. 差异化是一种瞄准广阔的大众市场，创造全行业可感知独特性的产品或服务的竞争战略。

Through the differentiation strategy, the firm produces non-standardized (that is, distinctive) products for customers who value differentiated features more than they value low cost. For example, superior product reliability and durability and high-performance sound systems are among the differentiated features of Toyota Motor corporation's Lexus products. However, Lexus offers its vehicles to customers at a competitive purchase price relative to other luxury automobiles. As with Lexus products, a product's unique attributes, rather than its purchase price, provide the value for which customers are willing to pay.

To maintain success with the differentiation strategy results, the firm must consistently upgrade differentiated features that customers value and/or create new valuable features (innovate) without significant cost increases. This approach requires firms to constantly change their product lines. These firms may also offer a portfolio of products that complement each other, thereby enriching the differentiation for the customer and perhaps satisfying a portfolio of consumer needs. Because a differentiated product satisfies customers' unique needs, firms following the differentiation strategy are able to charge premium price. The ability to sell a good or service at a price that substantially exceeds the cost of creating its differentiated features allows the firm to outperform rivals and earn above-average returns. Rather than costs, a firm using the differentiation strategy primarily concentrates on investing in and developing features that differentiate a product in ways that create value for customers. Overall, a firm using the differentiation strategy seeks to be different from its competitors on as dimensions as possible. The less similarity between a firm's goods or service and those of competitors, the more buffered it is from rivals' actions. Commonly recognized differentiated goods include Toyota's Lexus, Ralph Lauren's wide array of product lines, Caterpillar's heavy-duty earth-moving equipment, and McKinsey & Co. 's differentiated consulting services.

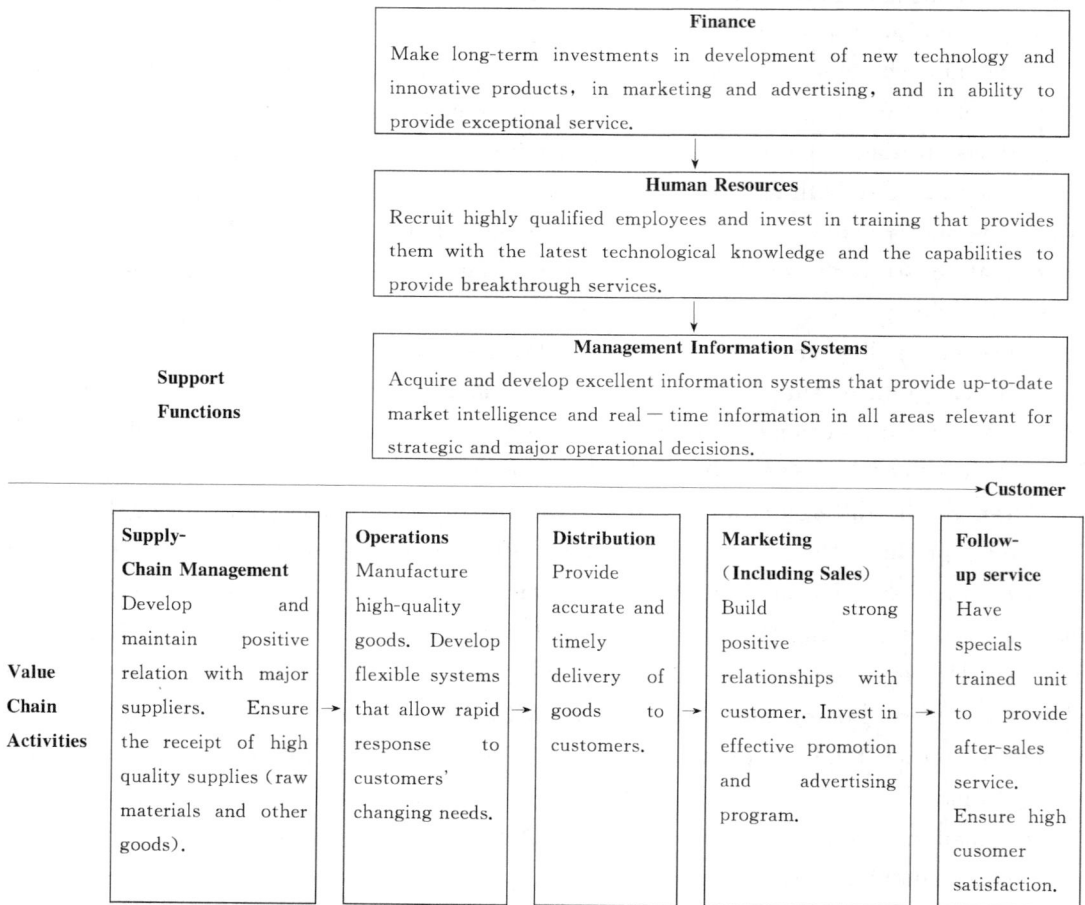

Finance
Make long-term investments in development of new technology and innovative products, in marketing and advertising, and in ability to provide exceptional service.

Human Resources
Recruit highly qualified employees and invest in training that provides them with the latest technological knowledge and the capabilities to provide breakthrough services.

Support Functions

Management Information Systems
Acquire and develop excellent information systems that provide up-to-date market intelligence and real—time information in all areas relevant for strategic and major operational decisions.

→Customer

Value Chain Activities

Supply-Chain Management	Operations	Distribution	Marketing (Including Sales)	Follow-up service
Develop and maintain positive relation with major suppliers. Ensure the receipt of high quality supplies (raw materials and other goods).	Manufacture high-quality goods. Develop flexible systems that allow rapid response to customers' changing needs.	Provide accurate and timely delivery of goods to customers.	Build strong positive relationships with customer. Invest in effective promotion and advertising program.	Have specials trained unit to provide after-sales service. Ensure high cusomer satisfaction.

Figure 5.2　Examples of Value-Creating Activities Associated with the Differentiation Strategy

5.6   Focus Strategies

The focus strategy is an integrated set of actions taken to produce goods or services that serve the needs of a particular competitive segment. Thus, firms use a focus strategy when they utilize their core competencies to serve the needs of a particular industry segment or niche to the exclusion of others. Examples of specific market segments that can be targeted by a focus strategy include a particular buyer group (e. g. , youths or senior citizens), a different segment of a product line (e. g. , products for professional painters or the do-it-yourself group), or a different geographic market (e. g. , northern or southern Italy by using a foreign subsidiary).

There are many specific customer needs firms can serve by using a focus strategy. For example, Los Angeles-based investment banking firm Greif & Company positions itself as "The Entrepreneur's Investment Bank. " Grief & Company is a leader in providing merger and acquisition advice to medium-sized businesses located in the western United States. Goya Foods is the largest U. S. -based Hispanic-owned food company in the United States. Segmenting the Hispanic market into unique groups, Goya offers more than 1600 products to consumers. The firm seeks "to be the be-all for the Latin community. " By successful using a focus strategy, firms such as these gain a competitive advantage in specific market niches or segments, even though they do not possess an industry-wide competitive advantage.

Although the breadth of a target is clearly a matter of degree, the essence of the focus strategy "is the exploitation of a narrow target's differences from the balance of the industry. " Firms using the focus strategy intend to serve a particular segment of an industry more effectively than can industry-wide competitors. They succeed when they effectively serve a segment whose unique needs are so specialized that broad-based competitors choose not to serve that segment or when they satisfy the needs of a segment being served poorly by industry-wide competitors.

When the lower cost and differentiation strategies have a broad (mass market) target, they are simply called cost leadership and differentiation. 当低成本战略和差异化战略具有广阔(大众市场)的目标时,则简单称之为成本领先和差异化。

When the lower cost and differentiation strategies are focused on a market niche (narrow target), however, they are called cost focus and differentiation focus. 当低成本战略和差异化战略专注于一个利基市场(狭窄目标)时,则称之为集中低成本和集中差异化。

Firms can create value for customers in specific and unique market segments by using the focused cost leadership strategy or the focused differentiation strategy.

A propitious niche is a company's specific competitive role that is so well suited to the firm's internal and external environment that other corporations are not likely to challenge or dislodge it. 适合利基是一家公司专有的竞争角色,与公司的内外部环境非常匹配,其他公司不太可能挑战或消除。

A firm's management must be always looking for strategic windows, that is, unique market opportunities available only for a limited time. 企业管理者必须一直努力寻找战略时机,即仅在有限时间内有效的独特市场机会。

Cost focus is a lower cost competitive strategy that focuses on a particular buyer group or geographic market and attempts to serve only this niche, to the exclusion of others. 集中低成本是一种低成本竞争战略,重点关注特定购买者群体或地域市场,并且试图只服务于这个利基市场,排斥其他市场细分。

Differentiation focus is a differentiation strategy that concentrates on a particular buyer group, product line segment, or geographic market. 集中差异化是一种差异化战略,重点关注特定购买者群体、产品线细分或地域市场。

5.7　Focused Cost Leadership Strategy

Based in Sweden, IKEA, a global furniture retailer with locations in 25 countries and territories and sales revenue of 21.1 billion euro's in 2008, uses the focused cost leadership strategy. Young buyers' desiring style at a low cost are IKEA's target customers. For these customers, the firm offers home furnishings that combine good design, function, and acceptable quality with low prices. According to the firm, "Low cost is always in focus. This applies to every phase of our activities."

IKEA emphasizes several activities to keep its costs low. For example, instead of relying primarily on third-party manufacturers, the firm's engineers design low-cost, modular furniture ready for assembly by customers. To eliminate the need for sales associates or decorators, IKEA positions the products in its stores so that customers can view different living combinations (complete with sofas, chairs, tables, etc.) in a single room like setting, which helps the customer imagine how furniture will look in the home. A third practice that helps keep IKEA's costs low is requiring customers to transport their own purchases rather than providing delivery service.

Although it is a cost leader, IKEA also offers some differentiated features that appeal to its target customers, including its unique furniture designs, in-store playrooms for children, wheelchairs for customer use, and extended hours. IKEA believes that these services and products "are uniquely aligned with the needs of customers, who are young, are not wealthy, are likely to have children, and, because they work, have a need to shop at odd hours." Thus, IKEA's focused cost leadership strategy also include some differentiated features with its low-cost products.

5.8　Integrated Cost Leadership/Differentiation Strategies

Most consumers have high expectations when purchasing a good or service. In general，it seems that most consumers want to pay a low price for products with somewhat highly differentiated features. Because of these customer expectations，a number of firms engage in primary value chain activities and support functions that allow them to simultaneously pursue low cost and differentiation. Firm seeking to do this use the integrated cost leadership/differentiation strategy. The objective of using this strategy is to efficiently produce products with some differentiated features. Efficient production is the source of maintaining low costs while differentiation is the source of creating unique value. Firms that successfully use the integrated cost leadership/differentiation strategy usually adapt quickly to new technologies and rapid changes in their external environments. Simultaneously concentrating on developing two sources of competitive advantage increase the number of primary and support activities in which the firm must become competent. Such firms often have strong networks with external parties that perform some of the primary and support activities. In turn，having skill in a larger number of activities makes a firm more flexible.

Porter argues that to be successful，a company or business unit must achieve one of the generic competitive strategies. Otherwise，the company or business unit is stuck in the middle of the competitive marketplace with no competitive advantage and is doomed to below-average performance. 波特认为,一个公司或业务单位要取得成功,必须采用一种通用竞争战略。否则,公司或业务单位会由于在竞争市场上没有竞争优势而卡在中间,注定只能具有平均水平的绩效表现。

Concentrating on the needs of its core customer group，Target Stores uses an integrated cost leadership/differentiation strategy as shown by its "Expect More. Pay Less' brand promise helped us to deliver greater convenience，increased savings and a more personalized shopping experience. " In 2010，Target remodeled 351 stores and provided a grater assortment of merchandise to include more grocery items and innovative products. It added more privately branded products to offer lower prices，created new mobile applications，and introduced distinctive Web strategies to continue to differentiate the services provided to customers.

Although each of Porter's generic competitive strategies may be used in any industry，In some instances certain strategies are more likely to succeed than others. In a fragmented industry，for example，in which many small and medium-size local companies compete for relatively small shares of the total market，focus strategies will likely predominate. 尽管波特的通用竞争战略可以适用于任何产业,但在实际中某些战略更容易成功。比如,在一个细

分的市场，一些小企业或中等规模的区域化企业利用聚焦化战略更容易在整个市场的相对小份额中获取市场地位。

As an industry matures，fragmentation is overcome and the industry tends to become a consolidated industry dominated by a few large companies. 当一个产业逐渐成熟和分化，那么这个产业就趋向于通过几家大的公司的控制来形成一个稳定的产业。

European-based Zara，which pioneered "cheap chic" in clothing apparel，is another firm using the integrated cost leadership/differentiation strategy. Zara offers current and desirable fashion goods at relatively low prices. To implement this strategy effectively requires sophisticated designers and means of managing costs，which fits Zara's capabilities. Zara can design and begin manufacturing a new fashion in three weeks，which suggests a highly flexible organization that can adapt easily to changes in the market or which competitors.

Vertical growth can be achieved by taking over a function previously provided by a supplier or distributor. 垂直增长通过接管以前由供应商或分销商所承担的职能来实现。

Vertical growth results in vertical integration，the degree to which a firm operates vertically in multiple locations on an industry's value chain from extracting raw materials to manufacturing to retailing. More specifically，assuming a function previously provided by a supplier is called backward integration. Assuming a function previously provide by a distributor is labeled forward integration. 垂直增长的结果是纵向一体化，是指一个公司在行业价值链上的多个节点垂直运作的程度，包括获取原材料、制造到零售。更具体地说，假设公司接管了以前由供应商承担的职能，称为后向一体化。假设公司接管了以前由分销商承担的职能，称为前向一体化。

Transaction cost economics proposes that vertical integration is more efficient than contracting for goods and services in the marketplace when the transaction costs of buying goods on the open market become too great. 交易成本经济学指出，当在公开市场上购买商品的交易成本过高的时候，纵向一体化比契约式商品和服务的提供方式更有效。

Horizontal Growth can be achieved by expanding the firm's products into other geographic locations and by increasing the range of products and services offered to current markets. Horizontal growth results in horizontal integration，the degree to which a firm operates in multiple locations at the same point in the industry's value chain. 水平增长通过将公司产品延伸到其他地域，以及扩大向目前市场所提供产品和服务的范围来实现。水平增长的结果是横向一体化，是指一个公司在多个地域的相同行业价值链节点上运作的程度。

When management realizes that the current industry is unattractive and that the firm lacks outstanding abilities or skills it could easily transfer to related products or services in other industries，the most likely strategy is conglomerate diversification-diversifying into an industry unrelated to its current one. 当管理层意识到目前行业缺乏吸引力，而公司又缺

乏出色的能力或技巧轻易地将相关产品或服务转移到其他行业时,最可能选择的战略是通过集团多元化进入与当前行业无关的行业。

A corporation may choose stability over growth by continuing its current activities without any significant change in direction. 一个公司可以通过选择稳定战略,也就是通过继续现在这种在方向上没有明显改变的活动来形成增长。

A pause/proceed-with-caution strategy is, in effect, a time-out-an opportunity to rest before continuing a growth or retrenchment strategy. 暂停或谨慎开始战略实际上继续采取增长或紧缩战略之前的暂停休息机会。

A no-change strategy is a decision to do nothing new-a choice to continue current operations and policies for the foreseeable future. 维持战略是不做任何新尝试的决策——在可预见的未来一段时间内维持目前的经营和政策。

A profit strategy is a decision to do nothing new in a worsening situation,but instead to act thought the company's problems are only temporary. 利润战略是在日益恶化的情况下不采取任何新举措的决策,而只解决公司的暂时性问题。

The profit strategy is an attempt to artificially support profits when a company's sales are declining by reducing investment and short-term discretionary expenditures. 利润战略是在公司销售额下降的时候,通过减少投资和短期任意支出,人为支持利润的尝试。

Management may pursue retrenchment strategies when the company has a weak competitive position in some or all of its product lines resulting in poor performance when sales are down and profits are becoming losses. 当一家公司在销售和利润上的表现比较差,在某些或所有产品线中的业绩的微弱优势逐渐丧失的时候,管理层可能会倾向于紧缩战略。

The turnaround strategy emphasizes the improvement of operational efficiency and is probably most appropriate when a corporation's problems are pervasive but not yet critical. 转型战略强调提高经营效率,当一家公司存在的问题很普遍但尚未构成威胁时,转型可能是最合适的选择。

The two basic phases of a turnaround strategy include contraction and consolidation. 转型战略包括消化和吸收两个基本阶段。

Contraction is the initial effort to quickly "stop the bleeding" with a general, across-the-board cutback in size and costs. 收缩是全线缩减规模和成本,有如快速止血般的努力。

The second phase, consolidation, is the implementation of a program to stabilize the now leaner corporation. 吸收是第二阶段,是为了稳定精简后的公司所实施的过程。

A captive company strategy is becoming another company's sole supplier or distributor in exchange for a long-term commitment from that company. The firm, in effect, gives up independence in exchange for security. 专属公司战略是指成为另一家公司的唯一供应商或分销商,交换条件是获得该公司的长期支持承诺。公司实际上是放弃了独立性来交换安全。

In a sell-out strategy, the entire company is sold. 出售战略是指出售整个公司。

If the corporation has multiple business lines, it may choose divestment, that is, the selling of a business unit. 如果一家公司拥有多个业务，可能选择剥离战略，即出售一个业务单位。

Bankruptcy involves giving up management of the firm to the courts in return for some settlement of the corporation's obligations. 破产是指放弃对公司的管理，交给法院处理公司债务。

In contrast to bankruptcy, which seeks to perpetuate the corporation, liquidation is piecemeal sale of all of the firm's assets. 与破产旨在延续公司相反，清算是将公司的全部资产进行拆分出售。

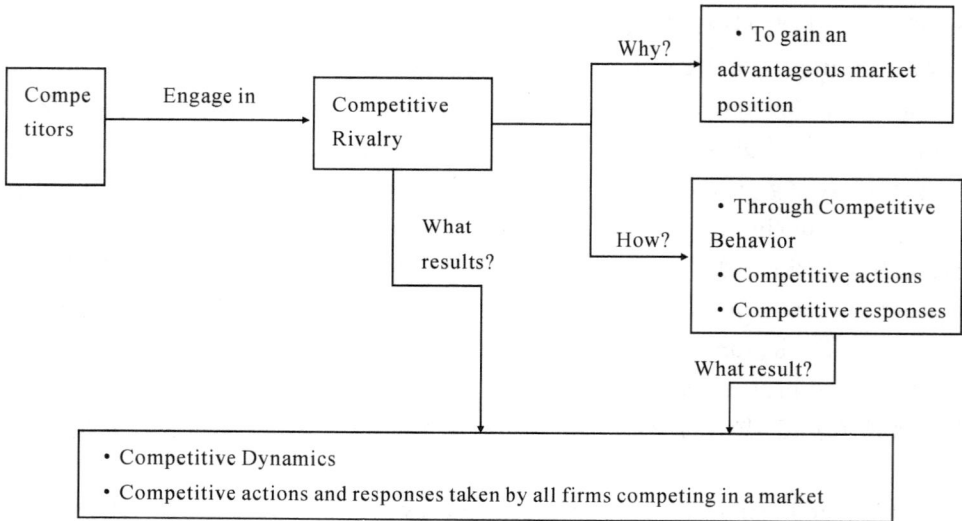

Figure 5.3 From Competitors to Competitive Dynamics

		Competitive Advantage: Lower Cost	Competitive Advantage: Differentiation
Competitive Scope	Broad Target	Cost Leadership	Differentiation
	Narrow Target	Cost Focus	Differentiation Focus

Figure 5.4 Porter's Generic Competitive Strategies

5.9   Levels of Diversification

Diversified firms vary according to their level of diversification and the connections between and among their businesses. The single- and dominant-business categories denote relatively low levels of diversification; more fully diversified firms are classified into related and unrelated categories. A firm is related through its diversification when its businesses share several links; for example, businesses may share products, technologies, or distribution channels. The more links among businesses, the more "constrained" is the relatedness of diversification. "Unrelated" refers to the absence of direct links between businesses.

5.10   Reasons for Diversification

A firm uses a corporate-level diversification strategy for a variety of reasons. Typically, a diversification strategy is used to increase the firm's value by improving its overall performance. Value is created either through related diversification or through unrelated diversification when the strategy allows a company's businesses to increase revenues or reduce costs while implementing their business-level strategies.

Other reasons for using a diversification strategy may have nothing to do with increasing the firm's value; in fact, diversification can have neutral effects or even reduce a firm's value. Value-neutral reasons for diversification include a desire to match and thereby neutralize a competitor's market power. Decisions to expand a firm's portfolio of businesses to reduce managerial risk can have a negative effect on the firm's value. Greater amounts of diversification reduce managerial risk in that if one of the businesses in a diversified firm fails, the top executive of that business does not risk total failure by the corporation. As such, this reduces the top executives' employment risk. In addition, because diversification can increase a firm's size and thus managerial compensation, managers have motives to diversify a firm to a level that reduce its value. Diversification rationales that may have a neutral or negative effect on the firm's value are discussed later in the chapter.

Operational relatedness and corporate relatedness ate two ways diversification strategies can create value. Studies of these independent relatedness dimensions show the importance of resources and key competencies. The Figure's vertical dimension depicts opportunities to share operational activities between businesses while the horizontal dimension suggests opportunities for transferring corporate-level core competencies. The firm with a strong capability in managing operational synergy, especially in sharing assets between its businesses, falls in the upper left quadrant, which also represents vertical sharing of assets through vertical integration. The lower right quadrant represents a highly developed corporate capability for transferring one or more core competencies across businesses.

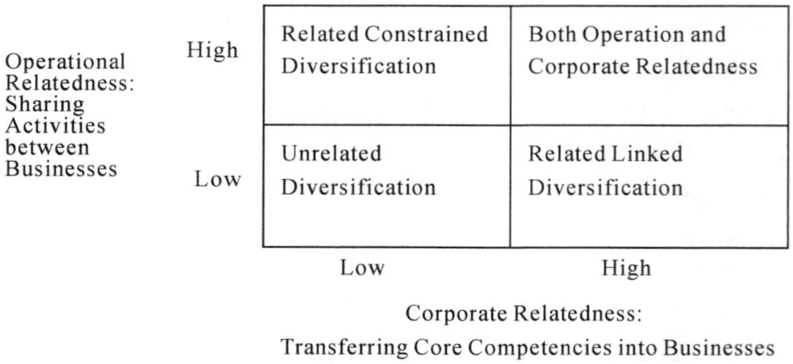

Operational Relatedness: Sharing Activities between Businesses	High	Related Constrained Diversification	Both Operation and Corporate Relatedness
	Low	Unrelated Diversification	Related Linked Diversification
		Low	High

Corporate Relatedness:
Transferring Core Competencies into Businesses

Figure 5. 5   Value-Creating Diversification Strategies: Operational and Corporate Relatedness

This capability is located primarily in the corporate headquarters office. Unrelated diversification is also illustrated in Figure 5. 5 in the lower left quadrant. Financial economies, rather than either operational or corporate relatedness, are the source of value creation for firms using the unrelated diversification strategy.

5. 11   Value-Creating Diversification: Related Constrained and Related Linked Diversification

With the related diversification corporate-level strategy, the firm builds upon or extends its resources and capabilities to build a competitive advantage by creating value for customers. The company using the related diversification strategy wants to develop and exploit economies of scope between its businesses. Available to companies operating in multiple product markets or industries, economies of scope are cost savings that the firm creates by successfully sharing some of its resources and capabilities or transferring one or more corporate-level core competencies that were developed in one of its businesses to another of its businesses.

As illustrated in Figure 5. 4, firms seek to create value from economies of scope through two basic kinds of operational economies: sharing activities and transferring corporate-level core competencies. The difference between sharing activities and transferring competencies is based on how separate resources are jointly used to create economies of scope. To create economies of scope tangible resources, such as plant and equipment or other business unit physical assets, often must be shared. Less tangible resources, such as manufacturing know-how transferred between separated activities with no physical or tangible resource involved is a transfer of a corporate-level core competence, not an operational sharing of activities.

Market stability is threatened by short product life cycles, short product design cycles, mew technologies, frequent entry by unexpected outsiders, repositioning by incumbents, and tactical redefinitions of market boundaries as diverse industries merge. 市场的稳定性正备受威胁:产品生命周期变短,产品设计周期变短,新技术层出不穷,未预期的外部企业频繁进入行业,现有企业重新定位,不同行业合并导致市场边界进行战术性重新定义。

One of the most popular aids to developing corporate strategy in a multi-business corporation is portfolio analysis. 在一家多元化公司的发展战略中,投资组合分析是最流行的目标之一。

In portfolio analysis, top management views its product lines and business units as a series of investments from which it expects a profitable return. 在投资组合分析过程中,高层管理者将各产品线和业务部门视为预期有收益的一系列投资。

The Boston Consulting Group (BCG) Growth-Share Matrix as depicted is the simplest way to portray a corporation's portfolio of investments. Each of the corporation's product lines or business units is plotted on the matrix according to the growth rate of the industry in which it competes, and its relative market share. 波士顿咨询集团(BCG)增长份额矩阵用最简单的方式描绘了一个公司的投资组合。公司的每条产品线或每个业务单位,根据所处行业的增长速度和相对市场份额绘入矩阵。

The GE Business Screen includes nine cells based on industry attractiveness, and business strength and competitive position. GE 业务荧屏基于行业吸引力、业务优势和竞争地位,分为 9 个单元。

How can portfolio analysis be used with strategic alliances? 投资组合分析如何用于战略联盟?

A study of 25 leading European corporations found four tasks of multi-alliance management that are necessary for successful alliance portfolio management: 25 家欧洲公司的一项研究发现,四项组合管理的任务对于成功的投资联盟管理是必要的:

1. Developing and implementing a portfolio strategy for each business unit and corporate policy for managing all the alliances of the entire company. 1. 为每个业务单位制定和实施投资组合战略和公司政策,用于管理整个企业的所有联盟

2. Monitoring the alliance portfolio in terms of implementing business unit strategies and corporate strategy and policies. 2. 从执行业务战略以及公司战略和政策等方面监控联盟的投资组合

3. Coordinating the portfolio to obtain synergies and avoid conflicts among alliances. 3. 协调投资组合,获得协同效应,避免联盟之间的冲突

4. Establishing an alliance management system to support other task of multi-alliance

management. 4.建立联盟管理系统,支持多联盟管理的其他任务

Corporate parenting, in contrast, views the corporation in terms of resources and capabilities that can be used to build business unit value as well as generate synergies across business units. 企业母合从资源和能力角度看待公司,这些资源和能力可用于建立业务单位价值,以及产生跨业务单位的协同效应。

Many people recommend that the search for appropriate corporate strategy involves three analytical steps: 许多人认为寻找合适的公司战略有三个分析步骤:

1. Examine each business unit(or target firm in the case of acquisition) in terms of its strategic factors. 按照自身的战略因素检查各业务单位(或收购案中的目标公司)

2. Examine each business unit(or target firm) in terms of area in which performance can be improved. 按照能够改善绩效的领域检查各业务单位(或目标公司)

3. Analyze how well the parent corporation fits with the business unit (or target firm). 分析母公司与业务单位(或目标公司)的匹配情况

When a horizontal strategy used to build synergy, it acts like a parenting strategy; when used to improve the competitive position of one or more business units, it can be thought of as a corporate competitive strategy. 当横向战略用于建立协同效应时,就是母合战略;当横向战略用来提高一个或多个业务单位的竞争地位时,则被认为是公司的竞争战略。

These multipoint competitors are firms that compete with each other not only in one business unit, but also in a number of business units. 这些多点竞争者不仅在一个业务单位与公司相互竞争,还在数个业务单位与公司相互竞争。

5.12 Classification of Enterprises Based on Capital Introduction Strategy

According to the proportion of operational resources, financial resources, shareholder investment resources, and shareholder residual resources in the total scale of liabilities and shareholder equity, enterprises can be divided into several types based on capital introduction strategies: operational driven by operational resources, debt financing driven by financial resources, shareholder driven by shareholder investment A profit driven model with residual resources as the main focus and a balanced utilization of various resources as the main focus. Of course, in many cases, enterprises will comprehensively utilize various resources to seek their own development. Obviously, different types of enterprise resource driven models demonstrate different resource driven strategies.

按照企业经营性资源、金融性资源、股东入资资源以及股东留剩资源在负债和股东权益总规模中的比重大小,可以将企业按照资本引入战略区分为几种类型:以经营性资源为主的经营驱动型、以金融性资源为主的债务融资驱动型、以股东入资为主的股东驱动型、以留剩

资源为主的利润驱动型以及以均衡利用各类资源为主的并重驱动型。当然,在很多情况下,企业会综合利用各类资源来谋求其自身的发展。显然,不同类型的企业资源驱动模式展示了不同的资源驱动战略。

Business driven enterprises that primarily rely on operational resources are often in a dominant position in the same industry competition, with a high proportion of operational liabilities to liabilities. The strategic connotation of such enterprises is very clear: to utilize their unique competitive advantages and maximize the use of upstream and downstream enterprise funds to support the operation and expansion of the enterprise. The strategic effect of business driven enterprises is: firstly, a large amount of funds required for business operation and expansion come from upstream and downstream enterprises without capital costs, thereby minimizing the financial costs of the enterprise; Secondly, to a considerable extent, solidify the business and financial connections between enterprises and upstream and downstream enterprises, making them an overall economic alliance; Thirdly, due to the inclusion of gross profit factors in the debt scale of prepayments, the actual debt scale of enterprises with high prepayments is higher than the calculated asset liability ratio; Fourthly, the high debt of the enterprise caused by this may not necessarily indicate high risk, but may instead reflect the competitive advantage of the enterprise. Of course, there is one exception. When a company's business activities lack market competitiveness, capital turnover is not flexible, and it is difficult to sustain, it will also show a situation of long-term high operating liabilities on the balance sheet. The formation of this financial situation cannot be considered as the result of the company's capital introduction strategy, but should be the result of serious operational difficulties.

以经营性资源为主的经营驱动型企业往往处于同行业竞争的主导性地位,经营性负债占负债的比例较高,这类企业的战略内涵十分清晰:利用自身独有的竞争优势,最大限度地占用上下游企业资金支撑企业的经营与扩张。经营驱动型企业的战略效应是:第一,企业经营与扩张所需资金大量来自于没有资金成本的上下游企业,从而最大限度地降低企业的财务成本;第二,在相当程度上固化企业与上下游企业的业务与财务联系,使其成为整体上的经济联盟体;第三,预收款项的负债规模由于包含了毛利因素,因而具有高预收款项企业的实际负债规模比计算出来的资产负债率要高;第四,由此引起的企业高负债不仅不一定表明企业的高风险,反而可能反映了企业的竞争优势。当然,有一种情况例外。当企业的经营活动缺乏市场竞争力、资金周转不灵、难以为继时,在资产负债表上也会表现为经营性负债长期居高不下的情形。此种财务状况的形成,就不能被认为是企业资本引入战略的结果,而应该是经营出现严重困难的结果。

Debt financing driven enterprises mainly relying on financial resources typically have a high proportion of financial liabilities in the total debt scale. This type of enterprise is often in a stage of rapid expansion, where shareholder investment and operational resources are difficult to meet the expansion funding needs. At this point, the funds required for the rapid development or expansion of enterprises can only be met through financial liabilities. Its strategic connotation is very clear: in a certain financing environment, to maximize the use of funds obtained from the company's financing capabilities to support its operation and expansion, enabling the company to achieve rapid development in a relatively short period of time. The strategic effect of debt financing driven enterprises is: firstly, a large amount of funds required for enterprise expansion come from upstream and downstream enterprises without capital costs, thereby maximizing the development speed of the enterprise; Secondly, due to the existence of certain capital cost factors in debt financing, the financial burden of enterprises will become an important consideration factor for the optimal financing structure; Thirdly, in order to reduce the impact of dynamic uncertainty in the financing environment, companies often encounter excessive financing problems.

以金融性资源为主的债务融资驱动型企业,其金融性负债通常占负债总规模的比重较高。这类企业往往处于企业快速扩张、股东入资和经营性资源难以满足扩张资金需求的发展阶段。此时,企业的快速发展或者扩张所需资金只能通过金融性负债来解决。其战略内涵十分清晰:在一定的融资环境下,最大限度地利用企业的融资能力所获得的资金来支持企业的经营与扩张,使企业能够在较短时间内实现快速发展。债务融资驱动型企业的战略效应是:第一,企业扩张所需资金大量来自于没有资金成本的上下游企业,从而最大限度地扩大企业的发展速度;第二,由于债务融资均存在一定的资本成本因素,因而企业的财务负担会成为最佳融资结构的重要考量因素;第三,为降低融资环境动态不确定性的影响,企业通常会出现过度融资问题。

Shareholders driven enterprises, mainly invested by shareholders, are often in the early stages of enterprise development. At this stage, enterprise debt financing activities and business activities are still difficult to bring funds for enterprise operation and development. The balance sheet shows that the scale of "Paid-in capital (or equity)" and "capital reserve" in shareholders' equity accounts for a high proportion of the sum of enterprise liabilities and shareholders' equity. It should be said that after a period of enterprise development, this situation will disappear. Of course, if a company's financial performance remains shareholder driven after a period of operation, it may mean that the company's product operations continue to fail to achieve ideal profits, the company's debt financing ability is low, or the company has not acted in debt financing. The strategic effect of shareholder driven enterprises is: firstly, in order to maintain the survival and development of the enterprise, the physical form of the assets invested by shareholders must meet the requirements of the enterprise's development strategy for the physical form of resources; Second, in the case of non cash capital contribution, the fairness of shareholders' valuation of capital contribution assets not only determines the Return on assets of the enterprise in the future, but also regulates the interest relationship between shareholders; Thirdly, the scale, physical form, and structure of assets invested by shareholders also significantly affect the governance structure and development direction of the enterprise.

以股东入资为主的股东驱动型企业往往处于企业发展的初级阶段。在该阶段，企业债务融资活动和经营活动还难以带来企业经营与发展所需资金，在资产负债表上的表现是：股东权益中的"实收资本（或者股本）"和"资本公积"这两个项目的规模占企业负债与股东权益之和的比重较高。应该说，在企业发展一段时期以后，这种情形就会消失。当然，如果企业经过一段时期的经营，其财务表现仍然是股东驱动型，则可能意味着该企业的产品经营持续不能获得理想利润，企业的债务融资能力较低，或者企业在债务融资方面没有作为。股东驱动型企业的战略效应是：第一，为了维持企业的生存与发展，股东对企业的入资资产的实物形态必须符合企业发展战略对资源实物形态的要求；第二，在非现金入资的情况下，股东用于入资资产估价的公允性既决定了企业未来资产的资产报酬率，还调节了股东间的利益关系；第三，股东入资资产的规模、实物形态及其结构还显著影响着企业的治理结构以及企业的发展方向。

Profit driven enterprises that primarily rely on remaining resources typically have a higher proportion of the sum of their surplus reserves and undistributed profits to the sum of their liabilities and shareholders' equity. This situation is often the result of the enterprise's development to a certain stage and accumulated considerable profits (at least the sum of its surplus reserve and undistributed profits is greater than the sum of Paid-in capital or capital stock and capital reserve). Essentially, using remaining resources to support the development of a company is equivalent to reinvesting shareholders in the company. Therefore, the strategic connotation of profit driven enterprise development is consistent with that of shareholder driven enterprise development strategy.

以留剩资源为主的利润驱动型企业，其盈余公积和未分配利润的规模之和占企业负债与股东权益之和的比重通常较高。这种情况的出现往往是企业发展到一定阶段并累积了相当规模的利润（至少其盈余公积和未分配利润的规模之和大于实收资本或者股本与资本公积之和）的结果。从本质上来说，用留剩资源支持企业的发展，等同于股东对企业的再投资。因此，利润驱动型企业发展的战略内涵与股东驱动型企业发展战略内涵是一致的。

Balanced utilization of various resources is a driving force for enterprises that comprehensively utilize various capital resources for development at any stage of enterprise development. In fact, most companies fall into this category. The difference between enterprises and between different stages of development is that there are significant differences in the contribution of different types of capital resources at different stages of enterprise development. Therefore, the strategic connotation of balanced utilization of various resources to drive the development of enterprises also varies with the differences in the contribution of different types of capital resources.

均衡利用各类资源的并重驱动型企业，是那些在企业发展的任一阶段，综合利用各种资本资源进行发展的企业。实际上,大多数企业属于此类情形。企业之间、企业在不同发展阶段之间的差异是:在企业发展的不同阶段,不同类型的资本资源的贡献度有着明显的差异。因此,均衡利用各类资源的并重驱动型企业发展的战略内涵也随着不同类型的资本资源贡献度的差异而不同。

Chapter 6　Functional Strategies and Strategic Choices

6.1　Operational Relatedness：Sharing Activities

Firms can create operational relatedness by sharing either a primary activity or a support activity. Firms using the related constrained diversification strategy share activities in order to create value. Procter & Gamble (P&G) uses this corporate-level strategy. P&G's paper towel business and baby diaper business both use paper products as a primary input to the manufacturing process. The firm's paper production plant produces inputs for both businesses and is an example of a shared activity. In addition，because they both produce consumer products，these two businesses are likely to share distribution channels and sales networks.

Functional strategy is the approach a functional area takes to achieve corporate and business unit objectives and strategies by maximizing resource productivity. 职能战略是职能部门充分利用资源实现公司和业务单位的目标和战略的一种方法。

Activity sharing is also risky because ties among a firm's businesses create links between outcomes. For instance，if demand for one business's product is reduced，it may not generate sufficient revenues to cover the fixed costs required to operate the shared facilities. These types of organizational difficulties can reduce activity-sharing success. Additionally，activity sharing requires careful coordination between the businesses involved. The coordination challengers must be managed effectively for the appropriate sharing of activities.

A tactic is a specific operating plan detailing how a strategy is to be implemented in terms of when and where it is to be put into action. 战术是具体的运营计划，它详细说明了战略如何实施，以及付诸行动的时间和地点。

A timing tactic deals with when a company implements a strategy. The first company to manufacture and sell a new product or service is called the first mover(or pioneer). 时间战术解决公司何时执行战略的问题。第一家制造和销售新产品或服务的公司称为先行者（或开拓者）。

Although activity sharing across businesses is not risk-free，research shows that it can create value. For example, studies of acquisitions of firms in the same industry，such as the banking industry and software，found that sharing resources and activities and thereby creating economies of scope contributed to post-acquisition increases in performance and higher returns to shareholders. Additionally, firms that sold off related units in which resource sharing was a possible source of economies of scope have been found to produce lower returns than those that sold off businesses unrelated to the firm's core business. Still other research discovered that firms with closely related businesses have lower risk. These results suggest that gaining economies of scope by sharing activities across a firm's businesses may be important in reducing risk and in creating value. Further, more attractive results are obtained through activity sharing when a strong corporate headquarters office facilitates it.

A market location tactic deals with where a company implements a strategy. 市场位置战术解决公司在哪里执行战略的问题。

A company or business unit can implement a competitive strategy either offensively or defensively. 一个公司或业务单位能够通过采取进攻战术或防御战术来实施竞争战略。

An offensive tactic attempts to take market share from an established competitor. 进攻战术就是试图从一个确定的竞争对手那里瓜分市场。

In offensive tactics, some of the methods used to attack a competitor's position are: frontal assault, flanking maneuver, encirclement, bypass attack, guerrilla warfare. 进攻战术中根据进攻的位置被用来描述为一些竞争方法,它们是正面进攻、侧面进攻、合围之势、迂回之术、游击战争。

According to Porter, defensive tactics aim to lower the probability of attack, divert attacks to less-threatening avenues, or lessen the intensity of an attack. 根据波特理论,防御战术的目标就是降低攻击的可能性,向威胁较少的道路上转移攻击或减少密集的攻击。

These defensive tactics that deliberately reduce short-term profitability to ensure long-term profitability, such as raise structural barriers, increase expected retaliation, Lower the inducement for attack. 提高结构性壁垒、提高预期的报复、降低攻击的诱导这些防御战术能通过减少短期盈利来确保长期的盈利能力。

6.2 Market Power

Firms using a related diversification strategy may gain market power when successfully using a related constrained or related linked strategy. Mark power exists when a firm is able to sell its products above the existing competitive level or to reduce the costs of its primary and support activities below the competitive level, or both. Mars' acquisition of the Wrigley assets was part of its related constrained diversification strategy and added market share to the Mars/Wrigley above Cadbury and Nestle, which had 10.1 and 7.7 percent of the market share, respectively, at the time and left Hershey with only 5.5 percent of the market.

In addition to efforts to gain scale as a means of increasing market power，as Mars did when it acquired Wrigley，firms can create market power through multipoint competition and vertical integration. Multipoint competition exists when two or more diversified firms simultaneously compete in the same product areas or geographic markets. The actions taken by UPS and FedEx in two markets，overnight delivery and ground shipping，illustrate multipoint competition. UPS has moved into overnight delivery，FedEx's stronghold；FedEx has been buying trucking and ground shipping assets to move into ground shipping，UPS's stronghold. Moreover，geographic competition for markets increases. The strongest shipping company in Europe is DHL. All three competitors are moving into large foreign markets to either gain a stake or to expand their existing share. If one of these firms successfully gains strong positions in several markets while competing against its rivals，its market power may increase. Interestingly，DHL had to exit the U. S. market because it was too difficult to compete against UPS and FedEx，which are dominant in the United States.

Marketing strategy deals with pricing，selling，and distributing a product. 营销战略处理产品的定价、销售和分销等方面的事宜。

Using a market development strategy，a company or business unit can（1）capture a larger share of an existing market for current products through market saturation and market penetration or（2）develop new use and/or markets for current products. 使用市场开发战略，公司或业务单位可以：（1）通过市场饱和/市场渗透为目前产品在已有市场获取更大的市场份额；（2）为现有产品开发新用途或市场。

6.3　Value-Neutral Diversification：Incentives and Resources

The objectives firms seek when using related diversification and unrelated diversification strategies all have the potential to help the firm create value by using a corporate-level strategy. However，these strategies，as well as single-and dominant-business diversification strategies，are sometimes used with value-neutral rather than value-creating objectives in mind. As we discuss next，different incentives to diversify sometimes exist，and the quality of the firm's resources may permit only diversification that is value neutral rather than value creating.

Cooperative strategies are those strategies that are used to gain competitive advantage within an industry by working with rather than against other firms. 合作战略是采取与其他公司合作而不是针锋相对的方式获得行业内竞争优势的战略。

A strategic alliance is a partnership of two or more corporations or business units formed to achieve strategically significant objectives that are mutually beneficial. 战略联盟是指两个或两个以上的企业或业务单位构成战略伙伴关系，旨在实现显著的双赢战略目标。

6.4　Resources and Diversification

As already discussed, firms may have several value-neutral incentives as well as value-creating incentives to diversify. However, even when incentives to diversify exist, a firm must have the types and levels of resources and capabilities needed to successfully use a corporate-level diversification strategy. Although both tangible and intangible resources facilitate diversification, they vary in their ability to create value. Indeed, the degree to which resources are valuable, rare, difficult to imitate, and non-substitutable influences a firm's ability to create value through diversification. For instance, free cash flows are a tangible financial resource that may be used to diversify the firm. However, compared with diversification that is grounded in intangible resources, diversification based on financial resources only is more visible to competitors and thus more imitable and less likely to create value on a long-term basis. Tangible resources usually include the plant and equipment necessary to produce a product and tend to be less-flexible assets. Any excess capacity often can be used only for closely related products, especially those requiring highly similar manufacturing technologies. For example, large computer makes such as Dell and Hewlett-Packard have underestimated the demand for tablet computers, especially-Apple's iPad. Apple developed the iPad and may expect it to eventually replace the personal computer. In fact, HP's and Dell's sales of their PCs have been declining since the introduction of the iPad. Apple expects to sell 70 million IPads in 2011 and analysts projects sales of the iPad to reach 246 million in 2014. HP and Dell likely need to diversify their product lines.

A mutual service consortium is a partnership of similar companies in similar industries who pool their resources to gain a benefit that is too expensive to develop alone, such as access to advanced technology. 双边服务联盟伙伴关系中,类似行业的类似公司联合双方资源获得独自开发费用太高的收益,比如获得先进技术。

Joint venture is a cooperative business activity, formed by two or more separate organizations for strategic purposes, that creates an independent business entity and allocates ownership, operational responsibilities, and financial risks and rewards to each member, while preserving their separate identity and autonomy. 合资企业是一种合作性业务活动。出于战略目的,两个或多个独立组织构建一个独立的商业实体,并在每个成员间分配所有权、经营责任、金融风险和回报,同时保留独立的身份和自主权。

Extremely popular in international undertaking because of financial and political-legal constraints, joint ventures are a convenient way for corporations to work together without losing their independence. 由于金融和政治法律方面的约束,合资企业在国际市场上极其流行,是一种公司之间共同合作又不丧失独立性的便捷方式。合资企业的缺点包括丧失控

制、利润降低、与合作伙伴产生冲突的可能性，以及技术优势向合作伙伴转移的可能性。

6.5　Value-Reducing Diversification: Managerial Motives to Diversify

Managerial motives to diversify can exist independent of value-neutral reasons and value-creating reasons. The desire for increased compensation and reduced managerial risk are two motives for top-level executives to diversify their firm beyond value-creating and value-neutral levels. In slightly different words, top-level executives may diversify a firm in order to diversify their own employment risk, as long as profitability does not suffer excessively.

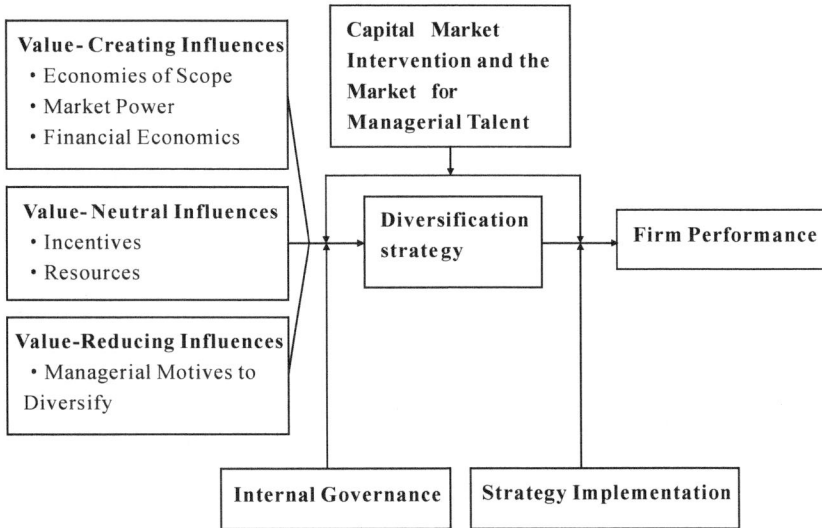

Figure 6.1　Summary Model of the Relationship Between Diversification and Performance

6.6　Learning and Developing New Capabilities

Firms sometimes complete acquisitions to gain access to capabilities they lack. For example, acquisitions may be used to acquire a special technological capability. Research shows that firms can broaden their knowledge base and reduce inertia of their capabilities when they acquire diverse talent through cross-border acquisitions. Of course, firms are better able to learn these capabilities if they share some similar properties with the firm's current capabilities. Thus, firms should seek to acquire companies with different but related and complementary capabilities in order to build their own knowledge base.

Using the product development strategy, a company or unit can（1）develop new products for existing markets or（2）develop mew products for new markets. 使用产品开

发战略,公司或业务单位可以:(1)为现有市场开发新产品;(2)为新市场开发新产品。

Many large food and consumer product companies in North America have followed a push strategy by spending a large amount of money on trade promotion in order to gain or hold shelf space in retail outlet. 北美许多大型食品和消费品公司遵循推动战略,在交易促进方面投入巨资,目的是增加或保持在零售店的铺货空间。

For new-product pioneers, skim pricing offers the opportunity to "skim the cream" from the top of the demand curve with a high price while the product is novel and competitors are few. 对于一些产品领先者来说,当产品新颖且竞争者较少时,撇脂定价以高价提供了"撇去奶油"的机会。

Penetration pricing, in contrast, use the experience curve to gain market share with a low price and then dominate the industry. 相反渗透定价用低价来获得市场份额,然后主导这个产业。

A number of large pharmaceutical firms are acquiring the ability to create "large molecule" drugs, also known as biological drugs, by buying biotechnology firms. Thus, these firms are seeking access to both the pipeline of possible drugs and the capabilities that these firms have to produce them. Such capabilities are important for large pharmaceutical firms because these biological drugs are more difficult to duplicate by chemistry alone. Biotech firms are focused on DNA research and have a biology base rather than a chemistry base. As an example, Sanofi-Aventis acquired Genzyme for $20 billion. Sanofi's hope is that the biotech company will help it keep rare-disease drugs in the pipeline without losing sales to more generic competition such as this that Sanofi keep expertise an genetics and biomarkers back to Sanofi. Such biomarkers "are biological substances in the body that help show the body is responding to disease and medication." If the acquisition is successful, there is added competitive advantage. Biological drugs must clear more regulatory barriers or hurdles which, when accomplished, add more to the advantage the acquiring firm develops through such acquisitions.

6.7　Three Basic Benefits of International Strategies

As noted, effectively using one or more international strategies can result in three basic benefits for the firm. These benefits facilitate the firm's effort to achieve strategic competitiveness when using an international strategy.

Firms can expand the size of their potential market-sometimes dramatically-by using an international strategy to establish stronger positions in markets outside their domestic market. As noted, access to additional consumers is a key reason Carrefour sees China as a major source of growth.

Takeda, a large Japanese pharmaceutical company, recently acquired Swiss drug maker Nycomed for $13. 7 billon. Buying Nycomed makes Takeda a major player in European markets. More significantly, the acquisition broadens Takeda's distribution capability in emerging markets "at a time when pharmaceutical firms world-wide are wrestling with the impact on revenue from the expiration of patents." In fact, the Nycomed deal will increase Takeda's sales in china about fourfold. Along with starbucks, Carrefour and Takeda are two additional companies relying on international strategy as the path to increased market size in china.

Firms such as Starbucks, Carrefour, and Takeda understand that effectively managing different consumer tastes and practices linked to cultural values or traditions in different markets is challenging. Nonetheless, they accept this challenge because of the potential to enhance the firm's performance. Other firms accept the challenge of successfully implementing an international strategy largely because of limited growth opportunities in their domestic market. This appears to be at least partly the case for major competitors Coca-Cola and PepsiCo, firms that have not been able to generate significant growth in their U. S. domestic markets for some time. Indeed, most of these firm's growth is occurring in international markets. These two firms approach international growth somewhat differently. PepsiCo, the world's largest snack-food maker as a result of its Frito-Lay division, relies "on chip sales overseas to make up for slower beverage sales volumes in North America." Less diversified than PepsiCo in terms of products but not in terms of geography, Coca-Cola is the world's largest producer of soft drink concentrates and syrups and the world's largest producer of Juice and Juice-related products. Selling its products in more than 200 countries, Coca-Cola derives only approximately 32 percent of its revenue from sales in North America as the cornerstone of its efforts to outperform PepsiCo, its chief rival.

6.8　International Business-Level Strategies

Firms considering the use of any international strategy first develop domestic-market strategies. One reason this is important is that the firm may be able to use some of the capabilities and core competencies it has developed in its domestic market as the foundation for competitive success in international markets. However, research results indicate that the value created by relying on capabilities and core competencies developed in domestic markets as a source of success in international markets diminishes as a firm's geographic diversity increases.

Firms do not select and then use strategies in isolation of market realities. In the case of international strategies, conditions in a firm's domestic market affect the degree to which the firm can build on capabilities and core competencies it established in that market to create capabilities and core competencies in international markets. The reason for this is grounded in Michael Porter's analysis of why some nations are more competitive relative to those industries in other nations. Porter's core argument is that conditions of factors in a firm's home base that is, in its domestic market-either hinder the firm's efforts to use an international business-level strategy for the purpose of establishing a competitive advantage in international markets or support those efforts. Porter identifies four factors as determinants of a national advantage that some countries possess. Interactions among these four factors influence a firm's choice of international business-level strategy.

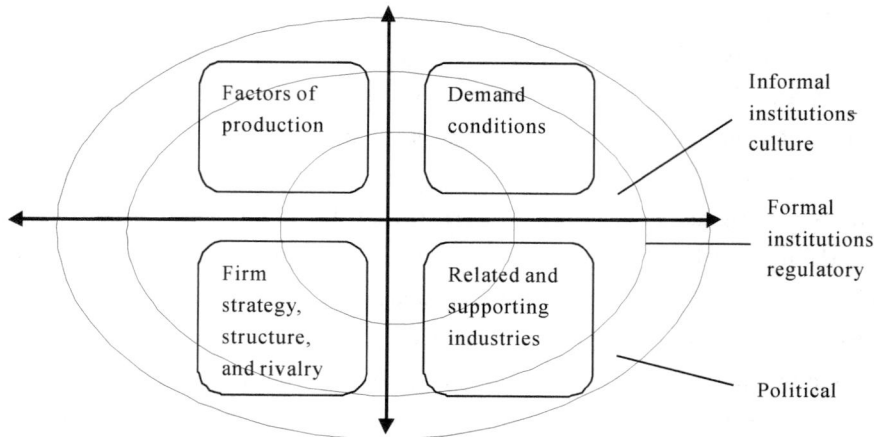

Figure 6.2 Determinants of National Advantage

6.9 Global Strategy

A global strategy is an international strategy in which a firm's home office determines the strategies business units are to use in each country or region. This strategy indicates that the firm has a high need for global integration and a low need for local responsiveness. These needs indicate that compared to a multi-domestic strategy, a global strategy seeks grater levels of standardization of products across country markets. The firm using a global strategy seeks to develop economies of scale as it produces the same or virtually the same products for distribution to customers throughout the world who are assumed to have similar needs. The global strategy offers greater opportunities to take innovations developed at the corporate level or in one market and apply them in other markets. Improvements in global accounting and financial reporting standards facilitate use of this strategy. A global is most effective for us when the differences between markets and the customers the firm is serving are insignificant.

6. 10 Acquisitions

When a firm acquires another company to enter an international market, it has completed a cross-border acquisition. Specifically, a cross-border acquisition is an entry mode through which a firm from one country acquires a stake in or purchases all of a firm located in another country acquires a stake in or purchases all of a firm located in another country.

As free trade expands in global markets, firms throughout the world are completing a large number of cross-border acquisitions. The ability of cross-border acquisitions to provide rapid access to new markets is a key reason for their growth. In fact, of the five entry modes, acquisitions often are the quickest means for firms to enter international markets.

Keys to Success in Achieving Low-Cost Leadership Scrutinize each cost-creating activity, identifying cost drivers Use knowledge about cost drivers to manage costs of each activity down year after year. Find ways to reengineer how activities are performed and coordinated—eliminate the costs of unnecessary work steps. Be creative in cutting low value-added activities out of value chain system—re-invent their industry value chain.

Financial strategy examines the financial implications of corporate and business level strategic options and identifies the best financial course of action. 财务战略检验公司和业务单位的战略选择对财务的影响,并确定行动的最佳财务过程。

The trade-off between achieving the desired debt-to-equity ratio and relying on internal long-term financing by way of cash flow is a key issue in financial strategy. 财务战略的关键问题是权衡实现既定债权比率,还是依赖现金流的内部长期融资方式。

In a leveraged buyout, a company is acquired in a transaction financed largely by debt, which is usually obtained from a third party such as an insurance company. Ultimately the debt is paid with money generated from the acquired company's operations or by sales of its assets. 在杠杆收购(LBO)中,在很大程度上,一个公司的收购交易通常依靠来自第三方的债务融资,如保险公司。最终的债务支付来自被收购公司的经营收入或资产出售。

Characteristics of a Low-Cost Provider Cost conscious corporate culture Employee participation in cost-control efforts. Ongoing efforts to benchmark costs. Intensive scrutiny of budget requests. Programs promoting continuous cost improvement Low-cost producers champion FRUGALITY while aggressively INVESTING in cost-saving improvements! Successful low-cost producers champion frugality but wisely and aggressively invest in cost-saving improvements!

When Does a Low-Cost Strategy Work Best? Price competition is vigorous Product is standardized or readily available from many suppliers. There are few ways to achieve differentiation that have value to buyers. Most buyers use product in same ways. Buyers incur low switching costs Buyers are large and have significant bargaining power Industry newcomers use introductory low prices to attract buyers and build customer base.

Pitfalls of Low-Cost Strategies Being overly aggressive in cutting price Low cost methods are easily imitated by rivals. Becoming too fixated on reducing costs and ignoring Buyer interest in additional features Declining buyer sensitivity to price. Changes in how the product is used Technological breakthroughs open up cost reductions for rivals.

Research and development (R&D) strategy deals with product and process innovation and improvement. One of the R&D choice is to be either a technological leader that pioneers an innovation or a technological follower that imitates the products of competitors. 研究与发展战略(简称研发战略)处理产品及流程创新和改进等事宜。研发选择包括作为开拓创新的技术领导者,或者作为模仿竞争对手产品的技术跟随者。

Differentiation Strategies Incorporate differentiating features that cause buyers to prefer firm's product or service over brands of rivals. Find ways to differentiate that create value for buyers and that are not easily matched or cheaply copied by rivals. Not spending more to achieve differentiation than the price premium that can be charged Objective Keys to Success.

Appeal of Differentiation Strategies A powerful competitive approach when uniqueness can be achieved in ways that Buyers perceive as valuable and are willing to pay for Rivals find hard to match or copy Can be incorporated at a cost well below the price premium that buyers will pay Which hat is unique?

Benefits of Successful Differentiation A product/service with unique and appealing attributes allows a firm to Command a premium price and/or Increase unit sales and/or Build brand loyalty＝Competitive Advantage.

Operations strategy determines how and where a product or service is to be manufactured，the level of vertical integration，the deployment of physical resources，and relationship with suppliers. 运营战略决定产品或服务的生产方式和位置、垂直一体化水平、物理资源调配以及与供应商的关系。

Increasing competitive intensity in many industries has forced companies to switch from traditional mass production using dedicated transfer lines to a continuous improvement production strategy，in which cross-functional work teams strive constantly to improve production processes. 在很多行业,日益提高的竞争强度迫使企业从传统的规模化生产转向使用专用传输线,采取持续改进的生产战略。在持续改进生产战略中,跨职能工作团队共同努力,持续改进生产流程。

In contrast to continuous improvement，mass customization requires flexibility and quick responsiveness. 与持续改进相比,大规模定制需要灵活性和快速反应能力。

Purchasing strategy deals with obtaining the raw materials, parts, and supplies needed to perform the operations function. 采购战略处理执行运营职能所需的获得原材料、零部件和供应等事宜。

Logistics strategy deals with the flow of products into and out of the manufacturing process. 物流战略处理制造过程中的产品输入和输出事宜。

Types of Differentiation Themes Unique taste—Dr. Pepper Multiple features—Microsoft Windows and Office Wide selection and one-stop shopping—Home Depot and Amazon. com. Superior service—FedEx, Ritz-Carlton Spare parts availability—Caterpillar More for your money—McDonald's, Wal-Mart Prestige—Rolex Quality manufacture—Honda, Toyota Technological leadership—3M Corporation, Intel Top-of-the-line image—Ralph Lauren, Chanel.

Sustaining Differentiation: The Key to Competitive Advantage. Most appealing approaches to differentiation Those hardest for rivals to match or imitate Those buyers will find most appealing Best choices for gaining a longer-lasting, more profitable competitive edge New product innovation. Technical superiority Product quality and reliability Comprehensive customer service Unique competitive capabilities.

Where to Find Differentiation Opportunities in the Value Chain Purchasing and procurement activities Product R&D and product design activities Production process/technology-related activities Manufacturing/production activities. Distribution-related activities Marketing, sales, and customer service activities Internally Performed Activities, Costs, &Margins Activities, Costs, &Margins of Suppliers Buyer/User Value Chains Activities, Costs, &Margins of Forward Channel Allies& Strategic Partners.

Human resource management(HRM) strategy attempts to find the best fit between people and the organization. 人力资源管理战略试图寻找员工与组织之间的最佳匹配方式。

Corporations are increasingly adopting information technology strategies to provide business units with competitive advantage. 越来越多的企业正在采用信息技术战略为业务单位提供竞争优势。

How to Achieve a Differentiation-Based Advantage Incorporate product features/attributes that lower buyer's overall costs of using product Approach 1Incorporate features/attributes that raise the performance a buyer gets out of the product. Approach Incorporate features/attributes that enhance buyer satisfaction in non-economic or intangible ways Compete on the basis of superior capabilities Approach.

Signaling Value as Well as Delivering Value Buyers seldom pay for value that is not perceived Signals of value may be as important as actual value when Nature of differentiation is hard to quantify Buyers are making first-time purchases Repurchase is infrequent Buyers are unsophisticated.

When Does a Differentiation Strategy Work Best? There are many ways to differentiate a product that have value and please customers Buyer needs and uses are diverse Few rivals are following a similar differentiation approach Technological change and product innovation are fast-paced.

Pitfalls of Differentiation Strategies Trying to differentiate on a feature buyers do not perceive as lowering their cost or enhancing their well-being Over-differentiating such that product features exceed buyers' needs Charging a price premium that buyers perceive is too high Failing to signal value Not understanding what buyers want or prefer and differentiating on the "wrong" things.

Outsourcing is purchasing from someone else a product or service that had been previously provided internally. 外包是从其他公司采购先前由公司内部提供的产品或服务。

Offshoring is the outsourcing of an activity or a function to a wholly owned company or an independent provider in another country. 离岸外包是将一项活动或一项职能外包给位于另一个国家的全资子公司或独立供应商。

Competitive Strategy Principle A low-cost producer strategy can defeat a differentiation strategy when buyers are satisfied with a standard product and do not see extra attributes as worth paying additional money to obtain! A low-cost provider strategy can defeat a differentiation strategy when buyers are satisfied with a standard product and do not see extra differentiating attributes as worth paying for!

Best Cost Provider Strategies Combine a strategic emphasis on low-cost with a strategic emphasis on differentiation Make an upscale product at a lower cost Give customers more value for the money Deliver superior value by meeting or exceeding buyer expectations on product attributes and beating their price expectations. Be the low-cost provider of a product with good-to-excellent product attributes, then use cost advantage to under price comparable brands Objectives.

Several strategies, which could be considered corporate, business, or functional, are very dangerous. Managers who have made a poor analysis or lack creativity may be trapped into considering them. 一些战略可能会被企业认为是十分危险的。那些不善于分析和缺乏创造力的管理者们可能会因为一些问题比如跟随领导者、再创本垒打、军备竞赛、全面出击、失控等而误入歧途。

How a Best-Cost Strategy Differs from a Low-Cost Strategy Aim of a low-cost strategy—Achieve lower costs than any other competitor in the industry. Intent of a best-cost strategy—Make a more upscale product at lower costs than the makers of other brands with comparable features and attributes. A best-cost provider cannot be the industry's absolute low-cost leader because of the added costs of incorporating the additional upscale features and attributes that the low-cost leader's product doesn't have.

Competitive Strength of a Best-Cost Provider Strategy. A best-cost provider's competitive advantage comes from matching close rivals on key product attributes and beating them on price Success depends on having the skills and capabilities to provide attractive performance and features at a lower cost than rivals. A best-cost producer can often out-compete both a low-cost provider and a differentiator when Standardized features/attributes won't meet the diverse needs of buyers Many buyers are price and value sensitive.

6.11   Unbundling and Outsourcing Strategies

De-Integration or unbundling involves narrowing the scope of the firm's operations, focusing on performing certain "core" value chain activities and relying on outsiders to perform the remaining value chain activities Concept Internally Performed Activities Suppliers Support Services Functional Activities Distributors or Retailers.

When Does Outsourcing Make Strategic Sense? Activity can be performed better or more cheaply by outside specialists. Activity is not crucial to achieve a sustainable competitive advantage. Risk exposure to changing technology and/or changing buyer preferences is reduced Operations are streamlined to Cut cycle time Speed decision-making Reduce coordination costs Firm can concentrate on doing those "core" value chain activities that best suit its resource strengths and capabilities.

Strategic Advantages of Outsourcing Improves firm's ability to obtain high quality and/or cheaper components or services. Improves firm's ability to innovate by interacting with "best-in-world" suppliers. Enhances firm's flexibility should customer needs and market conditions suddenly shift. Increases firm's ability to assemble diverse kinds of expertise speedily and efficiently. Allows firm to concentrate its resources on performing those activities internally which it can perform better than outsiders.

Pitfalls of Outsourcing Farming out too many or the wrong activities, thus Hollowing out its capabilities. Losing touch with activities and expertise that determine its overall long-term success. Offensive and Defensive Strategies Offensive Strategies Used to build new or stronger market position and/or create competitive advantage Defensive Strategies Used to protect competitive advantage(rarely are they the basis for creating advantage).

The Building and Eroding of Competitive Advantage competitive Buildup Period Benefit Period Erosion Period Strategic moves produce competitive advantage Moves by rivals erode competitive advantage Size of competitive advantage achieved Time. Competitive Strategy Principle Any competitive advantage currently held will eventually be eroded by the actions of competent, resourceful competitors!

Options for Mounting Strategic Offensives. Initiatives to match or exceed competitor strengths. Initiatives to capitalize on competitor weaknesses. Simultaneous initiatives on many fronts. End-run offensives. Guerrilla warfare tactics. Preemptive strikes.

Attacking Competitor Strengths Objectives Whittle away at a rival's competitive advantage. Gain market share by out-matching strengths of weaker rivals Challenging strong competitors with a lower price is foolhardy unless the aggressor has a cost advantage or advantage of greater financial strength!

Options for Attacking a Competitor's Strengths Offer equally good product at a lower price Develop low-cost edge，then use it to under-price rivals. Leapfrog into next-generation technologies Add appealing new features Run comparison ads Construct new plant capacity ahead of the rival or in the rival's market strong holds Offer a wider product line. Develop better customer service capabilities.

Attacking Competitor Weaknesses Objective Concentrate company strengths and resources directly against a rival's weaknesses. Go after Those customers a rival has that it is least equipped to serve Rivals providing sub-par customer service. Rivals with weaker marketing skills Geographic regions where rival is weak Segments rival is neglecting Weaknesses to Attack.

Launching Simultaneous Offensive son Many Fronts Launch several major initiatives to. Throw rivals off-balance Splinter their attention Force them to use substantial resources to defend their position. Objective. Appeal. A challenger with superior resources can overpower weaker rivals by out-competing them across-the-board long enough to become a market leader.

End-Run Offensives Objectives Dodge head-to-head confrontations that escalate competitive intensity or risk cut throat competition Attempt to maneuver around strong competitors—concentrate on areas of market where competition is weakest.

Optional Approaches for End-Run Offensives Introduce new products that redefine market and terms of competition Build presence in geographic areas where rival shave little presence. Create new segments by introducing products with different features to better meet buyer needs. Introduce next-generation technologies to leapfrog rivals.

Guerrilla Offenses Approach Use principles of surprise and hit-and-run to attack in locations and at times where conditions are most favorable to initiator Well-suited to small challengers with limited resources and market visibility Appeal.

Options for Guerrilla Offenses Make random，scattered raids on leaders' customers Occasional low-balling on price Intense bursts of promotional activity. Special campaigns to attract buyers from rivals plagued with a strike or having problems meeting delivery schedules. Challenge rivals encountering problems with quality，meeting delivery times，or providing adequate technical support. File legal actions charging antitrust violations，patent infringements，or unfair advertising. Preemptive Strikes Approach Involves moving first to secure an advantageous position that rivals are foreclosed or discouraged from duplicating!

Preemptive Strike Options Acquire firm which has exclusive control of a valuable technology Secure exclusive/dominant access to best distributors Tie up best or most sources of essential raw materials Secure best geographic locations Obtain business of prestigious customers. Expand capacity ahead of demand in hopes of discouraging rivals from following suit Build an image in buyers' minds that is unique or hard to copy. Choosing Who to Attack Four types of firms can be the target of an fresh offensive Market leaders Runner-up firms Struggling rivals on verge of going under. Small local or regional firms not doing a good job for their customers.

6.12   Offensive Strategies and Competitive Advantage

Strategic offensive offering strongest basis for competitive advantage usually entail. Developing lower-cost product design Making changes in production operations that lower costs or enhance differentiation Developing product features that deliver superior performance or lower users' costs. Giving more responsive customer service Escalating marketing effort Pioneering a new distribution channel Selling direct to end-users.

Offensive Strategy Principle The chances for a successful offensive initiative are improved when it is based on a company's resource strengths and strongest competencies and capabilities! Defensive Strategy Fortify firm's present position Help sustain any competitive advantage held Lessen risk of being attacked Blunt impact of any attack that occurs Influence challengers to aim attacks at other rivals Objectives. Defensive Strategies: Approaches Approach 1 Block avenues open to challengers Signal challengers that vigorous retaliations likely Approach.

Block Avenues Open to Challengers Participate in alternative technologies Introduce new features, add new models, or broaden product line to close gaps rivals may pursuer Maintain economy-priced models Increase warranty coverage. Offer free training and support services Reduce delivery times for spare parts Make early announcements about new products or price changes. Challenge quality or safety of rivals' products using legal tactics Sign exclusive agreements with distributors.

Signal Challengers Retaliation Is Likely Publicly announce management's strong commitment to maintain present market share Publicly announce plans to put adequate capacity in place to meet forecasted demand. Give out advance information about new products, technological breakthroughs, and other moves. Publicly commit firm to policy of matching prices and terms offered by rivals Maintain war chest of cash reserves. Make occasional counter-response to moves of weaker rival.

6.13　First-Mover Advantages

When to make a strategic move is often as crucial as what move to make. First-mover advantages arise when Pioneering helps build firm's image and reputation. Early commitments to new technologies, new-style components, and distribution channels can produce cost advantage. Loyalty of first time buyers is high Moving first can be a preemptive strike.

First-Mover Disadvantages Moving early can be a disadvantage (or fail to produce an advantage) when Costs of pioneering are sizable and loyalty of first time buyers is weak Innovator's products are primitive, not living up to buyer expectations Rapid technological change allows followers to leapfrog pioneers.

Timing and Competitive Advantage Principle 1Being a first-mover holds potential for competitive advantage in some cases but not in others Principle Being a fast follower can sometimes yield as good a result as being a first mover Principle Being a late-mover may or may not be fatal—it varies with the situation.

6.14　Customer Relationship Management Strategy

To be effective, data mining must occur within a context that allows an organization to change its behavior as a result of what it learns. It is no use knowing that wireless telephone customers who are on the wrong rate plan are likely to cancel their subscriptions if there is no one empowered to propose that they switch to a more appropriate plan as suggested in the sidebar. Data mining should be embedded in a corporate customer relationship strategy that spells out the actions to be taken as a result of what is learned through data mining. When low—value customers are identified, how will they be treated? Are there programs in place to stimulate their usage to increase their value? Or does it make more sense to lower the cost of serving them? If some channels consistently bring in more profitable customers, how can resources be shifted to those channels?

Data mining is a tool. As with any tool, it is not sufficient to understand how it works; it is necessary to understand how it will be used. Data mining, as we use the term, is the exploration and analysis of large quantities of data in order to discover meaningful patterns and rules. For the purposes of this book, we assume that the goal of data mining is to allow a corporation to improve its marketing, sales, and customer support operations through a better understanding of its customers. Keep in mind, however, that the data mining techniques and tools described here are equally applicable in fields ranging from law enforcement to radio astronomy, medicine, and industrial process control.

In fact, hardly any of the data mining algorithms were first invented with commercial applications in mind. The commercial data miner employs a grab bag of techniques borrowed from statistics, computer science, and machine learning research. The choice of a particular combination of techniques to apply in a particular situation depends on the nature of the data mining task, the nature of the available data, and the skills and preferences of the data miner.

Data mining is largely concerned with building models. A model is simply an algorithm or set of rules that connects a collection of inputs (often in the form of fields in a corporate database) to a particular target or outcome. Regression, neural networks, decision trees, and most of the other data mining techniques discussed in this book are techniques for creating models. Under the right circumstances, a model can result in insight by providing an explanation of how outcomes of particular interest, such as placing an order or failing to pay a bill, are related to and predicted by the available facts. Models are also used to produce scores. A score is a way of expressing the findings of a model in a single number. Scores can be used to sort a list of customers from most to least loyal or most to least likely to respond or most to least likely to default on a loan.

Classification deals with discrete outcomes: yes or no; measles, rubella, or chicken pox. Estimation deals with continuously valued outcomes. Given some input data, estimation comes up with a value for some unknown continuous variable such as income, height, or credit card balance. In practice, estimation is often used to perform a classification task. A credit card company wishing to sell advertising space in its billing envelopes to a ski boot manufacturer might build a classification model that put all of its cardholders into one of two classes, skier or non skier. Another approach is to build a model that assigns each cardholder a "propensity to ski score." This might be a value from 0 to 1 indicating the estimated probability that the cardholder is a skier. The classification task now comes down to establishing a threshold score. Anyone with a score greater than or equal to the threshold is classed as a skier, and anyone with a lower score is considered not to be a skier.

Prediction is the same as classification or estimation, except that the records are classified according to some predicted future behavior or estimated future value. In a prediction task, the only way to check the accuracy of the classification is to wait and see. The primary reason for treating prediction as a separate task from classification and estimation is that in predictive modeling there are additional issues regarding the temporal relationship of the input variables or predictors to the target variable.

Any of the techniques used for classification and estimation can be adapted for use in prediction by using training examples where the value of the variable to be predicted is already known, along with historical data for those examples. The historical data is used to build a model that explains the current observed behavior. When this model is applied to current inputs, the result is a prediction of future behavior.

6.15　Financial Strategy

Since the 21st century, economic globalization has undergone significant development, and strategic management of enterprises has become a very important part of enterprise management. With the research of many scholars, it has also developed well and formed a relatively complete financial strategy system. The main strategy of the enterprise includes the following aspects: product development, sales Financial resources, human resources, and infrastructure construction. It can be seen that enterprise strategy involves a wide range. Moreover, the strategies involved in an enterprise often need to be combined and used together. Therefore, many companies' financial strategies and policies to be implemented are composed of many aspects, and they need to cooperate with each other together. However, the strategic management of enterprises has indeed developed well, but the financial management of enterprises in strategic management has not received much attention from enterprises, which makes the research on financial strategy have some limitations.

21世纪以来,经济全球化得到了很大的发展,企业的战略管理也已经成为了企业管理中的一个很重要的部分,在很多学者的研究下得到了很好的发展,形成了比较完善的财务战略体系。企业主要的战略,包括以下几个方面:产品开发、销售.、财务资源,人力资源和基础建设等。可以看出企业战略所涉及的范围很广。并且,一个企业所涉及的策略往往都是需要组合在一起使用的。因此,很多企业的财务战略和要实行的政策,都是有很多方面组成的,他们在一起需要相互配合。但是,企业的战略管理确实发展的很好,但战略管理中的企业财务管理却没有受到企业的重视,这些轻视就使得财务战略的研究具有了一些局限性。

Therefore, with the development of China's economy and the further development of financial strategy theory, in order to further improve the strategic management of enterprises and make them more competitive, financial strategy should also be combined with research in other fields. From the perspective of enabling enterprises to achieve sufficient competitiveness and sustainable development, it is necessary to maximize the stimulation of enterprise goals and make financial strategic management choices, so as to enable enterprises to have their own development path in the highly competitive market, so that they can have their own healthy development. Financial strategy is of great significance for the sustainable development of an enterprise, and in-depth research on financial strategy is of great significance to the enterprise.

因此,随着我国经济的发展与财务战略理论的进一步发展,为了进一步完善企业的战略管理,使得企业有更大的竞争力,财务战略还应该要与其他领域的研究相结合。要从让企业获得足够的竞争力和持续发展的角度出发,把企业目标的最大化激发出来,来进行企业财务战略管理的选择,从而让企业在竞争力巨大的市场中有自己的发展之路,这样企业才有自己的良性发展。财务战略对一个企业的可持续发展意义重大,深入研究财务战略对企业有举足轻重的意义。

Financial strategy is based on the overall planning of the company's overall strategic objectives and value management，achieving the purpose of enterprise financial management objectives， as well as the strategic ideas， decision-making methods， management guidelines， and financial resources adopted by the company to optimize allocation as the standards for formulating financial strategies. Corporate strategy exists as a signboard for the overall strategic business of the enterprise, while financial strategy is only the backbone of the overall strategic business of the enterprise, and financial strategy is an important component of the overall strategy of the enterprise. Financial strategy is the support of an enterprise's overall business strategy and an indispensable component of the enterprise's overall business strategy. Only when an enterprise has formulated a good financial strategy can it maximize its capital gains，bring more economic benefits，and help to achieve the goal of maximizing financial Profit maximization.

财务战略是在公司总体战略目标的总体规划下,基于价值管理的基础上,实现企业财务管理目标的目的以及公司为优化配置而采用的战略思想,决策方法、管理指南和财务资源作为制定财务战略的标准。公司战略作为企业整体战略业务的招牌存在,财务战略只是企业整体战略业务的骨干部分,并且财务战略是企业整体战略的重要组成部分。财务战略是企业的整体经营战略的支撑,是企业整体经营战略不可或缺的组成部分。只有企业制定了良好的财务策略,才能使企业的资本收益最大化,带来更多的经济效益,有利于实现企业财务利润最大化的目标。

There are two methods for classifying corporate financial strategies. One is to divide them into investment strategies， fundraising strategies， operational strategies， and dividend strategies based on their functional types. The other type is divided into expansion type， robust type， defensive type， and contraction type financial strategies according to their comprehensive types. After classifying financial strategies，it is possible to have a clearer understanding of the types of financial strategies，which provides great convenience for the selection of financial strategies for enterprises in the future. It can also effectively understand the problems faced by enterprises and select the correct and suitable financial strategy for their own business type.

企业财务战略分类的方法有两种,一种是按职能类型分为投资战略、筹资战略、营运战略和股利战略。另一种是按综合类型分为扩展型、稳健型、防御型和收缩型财务战略。把财务战略分类以后,可以更清晰明了的知道财务战略的类型,为以后企业财务战略的选择提供了很大的方便,并且可以有效的了解企业所面临的问题,从而选择正确的、适合自己企业类型的财务战略。

Financial goal setting is the primary task of financial strategic analysis. Enterprises need to clarify their financial goals, such as increasing revenue, reducing costs, and improving profit margins. These goals need to be consistent with the company's strategic planning, laying the foundation for the future development of the enterprise. Financial risk management is one of the core contents of financial strategic analysis. Enterprises need to develop effective financial risk management strategies, including risk assessment, risk control, risk transfer, and risk monitoring. These strategies can help businesses cope with various risks and protect their financial security. The allocation of financial resources is one of the important contents of enterprise financial strategy analysis. Enterprises need to allocate financial resources reasonably, including funds, manpower, materials, etc., to meet the development needs of the enterprise. At the same time, enterprises also need to consider the sustainable development of financial resources to ensure their long-term competitiveness. Financial performance evaluation is one of the important links in financial strategic analysis. Enterprises need to evaluate their financial performance, including revenue, costs, profit margins, and other aspects. Through evaluation, companies can understand their financial situation, identify problems, and make timely adjustments. Financial information disclosure is one of the important contents of enterprise financial strategy analysis. Enterprises need to timely and accurately disclose their financial information, including financial statements, financial indicators, etc. This information can help investors and stakeholders understand the financial situation of a company, improve its transparency and credibility. The above is the main content of financial strategy analysis. When formulating financial strategies, enterprises need to fully consider these contents to ensure the scientific and effective financial management. At the same time, enterprises also need to develop financial strategies that are in line with their own actual situation to promote the healthy development of the enterprise.

财务目标设定是财务战略分析的首要任务。企业需要明确自己的财务目标,比如增加收入、降低成本、提高利润率等。这些目标需要与公司的战略规划相一致,为企业未来的发展奠定基础。财务风险管理是财务战略分析的核心内容之一。企业需要制定有效的财务风险管理策略,包括风险评估、风险控制、风险转移和风险监控等。这些策略可以帮助企业应对各种风险,保护企业的财务安全。财务资源配置是企业财务战略分析的重要内容之一。企业需要合理配置财务资源,包括资金、人力、物资等,以满足企业的发展需求。同时,企业还需要考虑财务资源的可持续发展,确保企业的长期竞争力。财务绩效评估是财务战略分析的重要环节之一。企业需要对自己的财务绩效进行评估,包括收入、成本、利润率等方面。通过评估,企业可以了解自己的财务状况,发现问题并及时调整。财务信息披露是企业财务战略分析的重要内容之一。企业需要及时、准确地披露自己的财务信息,包括财务报表、财务指标等。这些信息可以帮助投资者和利益相关者了解企业的财务状况,提高企业的透明度和信誉度。企业在制定财务战略时,需要充分考虑这些内容,以确保财务管理的科学性和有效性。同时,企业还需要根据自身的实际情况,制定符合自己的财务战略,以推动企业的健康发展。

With the further development of economic globalization, the connections between each enterprise continue to increase, so the internal and external environment in which each enterprise is located has become more complex, and risks have also continued to increase. However, there are still many companies that have not paid attention to the long-term impact of financial strategic management on the enterprise. If enterprises want to survive in a highly competitive market, they must attach importance to the position of financial strategy in enterprise management. So more and more scholars have conducted research on the financial strategy of enterprises and proposed many of their own understandings. From the research of foreign scholars, it can be seen that corporate financial strategy can serve as a set of strategies for the development of enterprises, and it is a plan for the future of enterprises, providing a path for their development. From the perspective of scholars both domestically and internationally, each scholar has emphasized the importance of financial strategy for enterprises. Therefore, the issue of financial strategy in enterprises is currently a very important issue, and the importance of financial strategy in the world has also been formed. In a world economy with such tremendous competitive pressure, being able to choose the right financial strategy for a company means gaining more control and adding a chip in the competition with other companies. So we need to use the correct financial strategy analysis method to select the appropriate financial strategy.

随着经济全球化的进一步发展,企业之间的联系不断增加,所以每个企业所处的内外部环境都变得更加复杂,风险也随之不断增加。但是,还有很多企业没有关注到财务战略管理对企业持续长远的作用。企业要想在竞争力巨大的市场中生存下去,就必须重视起财务战略在企业中管理的地位。所以越来越多的学者都对企业的财务战略进行了研究,并提出了很多自己的理解。从国外学者的研究来看,企业财务战略可以作业企业发展的一套策略,是企业未来的一种规划,为公司的发展提供了一条道路。综合国内外学者的观点来看,每个学者都强调出了财务战略对于企业的重要性,所以目前在企业中的财务战略的问题,是一个十分重要的问题,世界对于财务战略的重要性也已经形成了。在竞争压力如此巨大的的世界经济体中,企业能够选择出正确的财务战略,就意味着在与其他企业的竞争中多了一分把握,增添了一分筹码。所以我们要用正确的财务战略分析方法选择出合适的财务战略。

Chapter 7　Strategy Implementation

7.1　Implementing Internal Innovations

An entrepreneurial mind-set is required to be innovative and to develop successful internal corporate ventures. Because of environmental and market uncertainty, individuals and firms must be willing to take risks to commercialize innovations. Although they must continuously attempt to identify opportunities, they must also select and pursue the best opportunities and do so with discipline. Employing an entrepreneurial mind-set entails not only developing new products and markets but also execution in order to do these things effectively. Often, firms provide incentives to managers to be entrepreneurial and to commercialize innovations.

Strategy implementation is the sum total of the activities and choices required for the execution of a strategic plan. It is the process by which strategies and policies are put into action through the development of programs, budgets, and procedures. 战略实施是执行一项战略计划所需要的活动和选择的总和。战略实施是通过开发各种方案、预算和流程,将战略和政策付诸行动的过程。

Having processes and structures in place through which a firm can successfully implement the outcomes of internal corporate ventures and commercialize the innovations is critical. Indeed, as the Strategic Focus on 3M illustrates, the successful introduction of innovations into the marketplace reflects implementation effectiveness. In the context of internal corporate ventures, managers must allocate resources, coordinate activities, communicate with many different parties in the organization, and make a series of decisions to convert the innovations resulting from either autonomous or induced strategic behaviors into successful market entries. Organizational structures are the sets of formal relationships that support processes managers use to commercialize innovations.

Effective integration of the various functions involved in innovation processes from engineering to manufacturing and, ultimately, market distribution is required to implement the incremental and radical innovations resulting from internal corporate ventures. Increasingly, product development teams are being used to integrate the activities associated with different organizational functions. Such integration involves coordinating and applying the knowledge and skills of different functional areas in order to maximize innovation. Teams must help to make decisions as to which projects should be commercialized and which ones should end. Although ending a project is difficult, sometimes because of emotional commitments to innovation-based projects, effective teams recognize when conditions change such that the innovation cannot create value as originally anticipated.

A program is a statement of the activities or steps needed to accomplish a single-use plan. The purpose of a program is to make the strategy action-oriented. 方案是对完成一个单独用途的计划所需活动或步骤的详细说明。方案的目的是确保战略立足于行动。

Procedures, sometimes termed standard operating procedures(SOPs), are a system of sequential steps or techniques that describe in detail how a particular task or job is to be done. 流程有时称为标准作业流程(SOPs)，是一个系统，详细描述完成一项特定任务或工作的连续步骤或方法。

7.2   Facilitating Integration and Innovation

Shared values and effective leadership are important for achieving cross-functional integration and implementing innovation. Highly effective shared values are framed around the firm's vision and mission and become the glue that promotes integration between functional units. Thus, the firm's culture promotes unity and internal innovation.

Strategic leadership is also highly important for achieving cross-functional integration and promoting innovation. Leaders set the goals and allocate resources. The goals include integrated development and commercialization of new goods and services. Effective strategic leaders also ensure a high-quality communication system to facilitate cross-functional integration. A critical benefit of effective communication is the sharing of knowledge among team members. Effective communication thus helps create synergy and gains team members' commitment to an innovation throughout the organization. Shared values and leadership practices shape the communication systems that are formed to support the development and commercialization of new products.

7.3   Organizational Structure and Controls

Research shows that organizational structure and the controls that are a part of the structure affect firm performance. In particular, evidence suggests that performance declines when the firm's strategy is not matched with the most appropriate structure and controls. Even though mismatches between strategy and structure do occur, research indicates that managers try to act rationally when forming of changing their firm's structure. His record of success at General Electric suggests that CEO Jeffrey Immelt pays close attention to the need to make certain that strategy and structure remain matched, as evidenced by restructuring alignments in GE Capital, GE's financial service group, during the economic downturn.

Organization structure specifies the firm's reporting relationships, procedures, controls, and authority and decision-making processes. Developing an organizational structure that effectively supports the firm's strategy is difficult, especially because of the uncertainty about cause-effect relationships in the global economy's rapidly changing and dynamic competitive environments. When a structure's elements are properly aligned with one another, the structure facilitates effective use of the firm's strategies. Thus, organizational structure is a critical component of effective strategy implementation processes.

A firm's structure specifies the work to be done and how to do it, given the firm's strategy or strategies. Thus, organizational structure influences how managers work and the decision resulting from that work. Supporting the implementation of strategies, structure is concerned with processes used to complete organizational tasks. Having the right structure and process is important. For example, many product-oriented firms have been moving to develop service businesses associated with those products. This strategy has been used by GE. However, research suggests that developing a separate division for such services in product-oriented companies, rather than managing the service business within the product divisions, leads to additional growth and profitability in the service business. GE developed a separate division for its financial services businesses and this helped facilitate GE's growth over the last two decades.

Organizational controls are an important aspect of structure. Organizational controls guide the use of strategy, indicate how to compare actual results with expected results, and suggest corrective actions to take when the difference is unacceptable. When fewer differences separate actual from expected outcomes, the organization's controls are more effective. It is difficult for the company to successfully exploit its competitive advantages without effective organizational controls. Properly designed organizational controls provide clear insights regarding behaviors that enhance firm performance. Firms use both strategic controls and financial controls to support the implementation and use of their strategies.

Strategic controls are largely subjective criteria intended to verify that the firm is using appropriate strategies for the conditions in the external environment and the company's competitive advantages. Thus, strategic controls are concerned with examining the fit between what the firm might do and what it can do. Effective strategic controls help the firm understand what it takes to be successful. Strategic controls demand rich communications between managers responsible for using them to judge the firm's performance and those with primary responsibility for implementing the firm's strategies. These frequent exchanges are both formal and informal in nature.

Strategic controls are also used to evaluate the degree to which the firm focuses on the requirements to implement its strategies. For a business-level strategy, for example, the strategic controls are used to study primary and support activities to verify that the critical activities are being emphasized and properly executed. In fact, Nokia failed to employ effective strategic controls and is now fighting for survival as a result. With related corporate-level strategies, strategic controls are used by corporate strategic leaders to verify the sharing of appropriate strategic factors such as knowledge, markets, and technologies across businesses. To effectively use strategic controls when evaluating related diversification strategies, headquarter executives must have a deep understanding of each unit's business-level strategy. As we described in the Opening Case, Borders' significant strategic problems likely stemmed at least partly from the ineffective use of strategic controls.

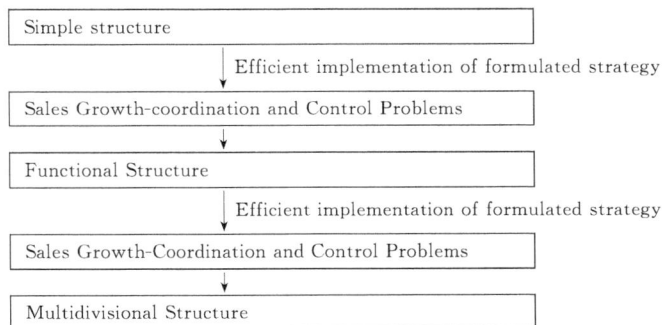

Simple structure
↓ Efficient implementation of formulated strategy
Sales Growth-coordination and Control Problems
↓
Functional Structure
↓ Efficient implementation of formulated strategy
Sales Growth-Coordination and Control Problems
↓
Multidivisional Structure

Figure 7.1   Strategy and Structure Growth Pattern

7.4   Functional Structure

The functional structure consists of a chief executive officer and a limited corporate staff, with functional line managers in dominant organizational areas such as production, accounting, marketing, R&D, engineering, and human resources. This structure allows for functional specialization, thereby facilitating active sharing of knowledge within each functional area. Knowledge sharing facilitates career paths as well as professional development of functional specialists. However, a functional orientation can negatively affect communication and coordination among those representing different organizational functions. For the reason, the CEO must verify that the decisions and actions of individual business functions promote the entire firm rather than a single function. The functional structure supports implementing business-level strategies with low levels of diversification. When changing from a simple to a functional structure, firms want to avoid introducing value-destroying bureaucratic procedures such as failing to promote innovation and creativity.

Matches between Corporate-Level Strategies and the Multidivisional Structure

As explained earlier, Chandler's research shows that the firm's continuing success leads to product or market diversification or both. The firm's level of diversification is a function of decisions about the number and type of businesses in which it will compete as well as how it will manage the businesses. Geared to managing individual organizational functions, increasing diversification eventually creates information processing, coordination, and control problems that the functional structure cannot handle. Thus, using a diversification strategy requires the firm to change from the functional structure to the multidivisional structure to develop an appropriate strategy/structure match.

Corporate-level strategies have different degrees of product and market diversification. The demands created by different levels of diversification highlight the need for a unique organizational structure to effectively implement each strategy (see Figure 7.2).

Cisco must use a differentiation strategy in order to compete in its several high technology product market segments. However, given the presence of major competitors in those markets, such as Hewlett-Packard and Huawei, and its loss of market share in its core market of routers, Cisco must also be sensitive to costs. Thus, the horizontal structure can be useful to integrate the two disparate dimensions of structure needed to implement Cisco's integrated cost leadership-differentiation strategy. In addition, Cisco needs to coordinate several related product units, and the horizontal structure should facilitate this cooperation. Therefore, Cisco's approach is similar to the cooperative M-form structure, discussed next.

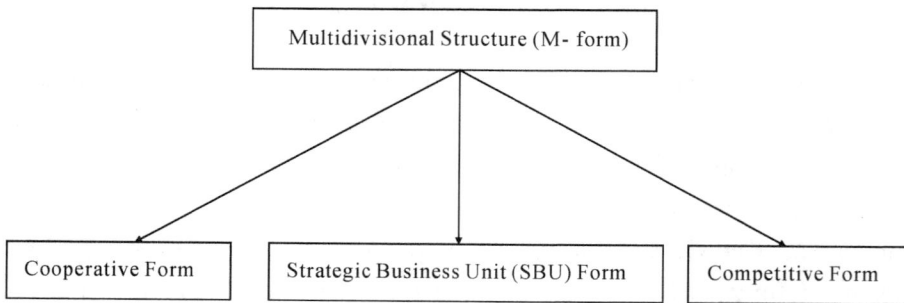

Figure 7.2   Three Variations of the Multidivisional Structure

Using the Strategic Business Unit Form of the Multidivisional Structure to implement the Related Linked Strategy

Firms with fewer links or less constrained links among their divisions use the related linked diversification strategy. The strategic business unit form of the multidivisional structure supports implementation of this strategy. The strategic business unit (SBU) form is an M-form structure consisting of three levels: corporate headquarters, strategic business units, and SBU divisions. The SBU structure is used by large firms and can be complex, given associated organization size and product and market diversity.

The divisions within each SBU are related in terms of shared products or markets or both, but the divisions of one SBU have little in common with the divisions of the other SBUs. Divisions within each SBU share product or market competencies to develop economies of scope and possibly economies of scale. The integrating mechanisms use by the divisions in this structure can be equally well used by the divisions within the individual strategic business units that are part of the SBU form of the multidivisional structure. In this structure, each SBU is a profit center that is controlled and evaluated by the headquarters office. Although both financial and strategic controls are important, on a relative basis financial controls are vital to headquarters' evaluation of each SBU; strategic controls are critical when the heads of SBUs evaluate their divisions' performances. Strategic controls are also critical to the headquarters' efforts to determine whether the company has formed an effective portfolio of businesses and whether those businesses are being successfully managed. Therefore, there is need for strategic structures that promote exploration to identify new products and markets, but also for actions that exploit the current product lines and markets.

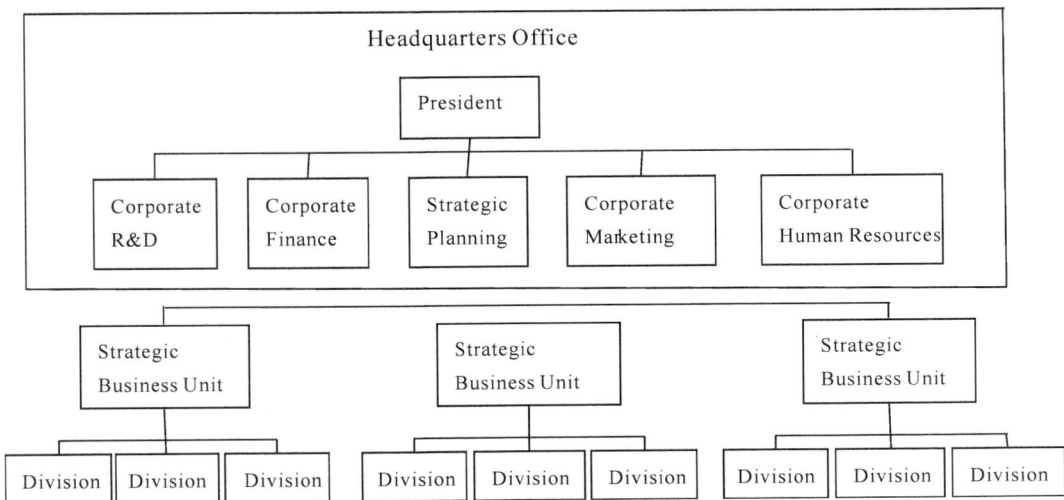

Figure 7.3   SUB Form of the Multidivisional Structure for Implementing a Related Linked Strategy

7.5   The Role of Top-Level Managers

Top-level managers play a critical role in that they are charged to make certain their firm is able to effectively formulate and implement strategies. Top-level managers' strategic decisions influence how the firm is designed and goals will be achieved. Thus, a critical element of organizational success is having a top management team with superior managerial skills.

Managers often use their discretion when making strategic decisions, including those concerned with effectively implementing strategies. Managerial discretion differs significantly across industries. The primary factors that determine the amount of decision-making discretion held by a manager are (1) external environmental sources such as the industry structure, the rate of market growth in the firm's primary industry, and the degree to which products can be differentiated; (2) characteristics of the organization, including its size, age, resources, and culture; and (3) characteristics of the manager, including commitment to the firm and its strategic outcomes, tolerance for ambiguity, skills in working with different people, and aspiration levels. Because strategic leaders' decisions are intended to help the firm gain a competitive advantage, how managers exercise discretion when determining appropriate strategic actions is critical to the firm's success.

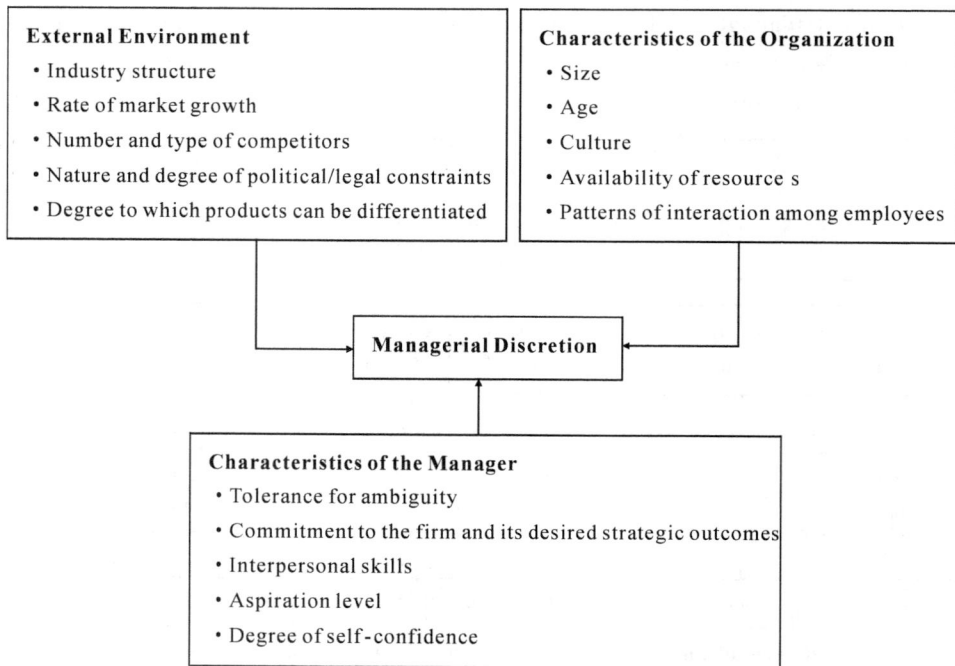

External Environment
- Industry structure
- Rate of market growth
- Number and type of competitors
- Nature and degree of political/legal constraints
- Degree to which products can be differentiated

Characteristics of the Organization
- Size
- Age
- Culture
- Availability of resource s
- Patterns of interaction among employees

Managerial Discretion

Characteristics of the Manager
- Tolerance for ambiguity
- Commitment to the firm and its desired strategic outcomes
- Interpersonal skills
- Aspiration level
- Degree of self-confidence

Figure 7. 4   Factors Affecting Managerial Discretion

In addition to determining new strategic initiatives, top-level managers develop a firm's organizational structure and reward systems. Top executives also have a major effect on a firm's culture. Evidence suggests that managers' values are critical in shaping a firm's cultural values. Accordingly, top-level managers have an important effect on organizational activities and performance. Because of the challenges top executives face, they often are more effective when they operate as top management teams.

7.6   Top Management Teams

In most firms, the complexity of challenges and the need for substantial amounts of information and knowledge require strategic leadership by a team of executives. Using a team to make strategic decisions also helps to avoid another potential problem when these decisions are made by the CEO alone: managerial hubris. Research evidence shows that when CEOs begin to believe glowing press accounts and to feel that they are unlikely to make errors, they are more likely to make poor strategic decisions. Top executives need to have self-confidence but must guard against allowing it to become arrogance and a false belief in their own invincibility. To guard against CEO overconfidence and poor strategic decisions, firms often use the top management team to consider strategic opportunities and problems and to make strategic decisions. The top management team is composed of the key individuals who are responsible for selecting and implementing the firm's strategies. Typically, the top management team includes the officers of the corporation, defined by the title of vice president and above or by service as a member of the board of directors. The quality of the strategic decisions made by a top management team affects the firm's ability to innovate and engage in effective strategic change.

7.7   Managerial Succession

The choice of top executives-especially CEOs is a critical decision with important implications for the firm's performance. Many companies use leadership screening systems to identify individuals with managerial and strategic leadership potential as well as to determine the criteria individuals should satisfy to be candidates for the CEO position.

The most effective of these systems assesses people within the firm and gains valuable information about the capabilities of other companies' managers, particularly their strategic leaders. Based on the results of these assessments, training and development programs are provided for current individuals in an attempt to preselect and shape the skills of people who may become tomorrow's leaders. Because of the quality of its programs, General Electric "is famous for developing leaders who are dedicated to turning imaginative ideas into leading products and services." However, there are many companies that do not have succession plans for their top executives. For example, a recent survey found that 43 percent of the largest public companies in the United States had no formal succession plan for their CEOs. Of those companies with plans, only about 20 percent were satisfied with their succession processes.

Synergy can take place in one of six ways: shared know-how, coordinated strategies, shared tangible resources, economies of scale or scope, pooled negotiating power, and new business creation. 产生协同效应的方式有六种:共享知识、协调战略、共享有形资源、规模经济或范围经济、集中谈判力量以及创造新业务。

7.8　Sustaining an Effective Organizational Culture

We defined organizational culture as a complex set of ideologies, symbols, and core values that are shared throughout the firm and influence the way business is conducted. Evidence suggests that a firm can develop core competencies in terms of both the capabilities it possesses and the way the capabilities are leveraged when implementing strategies to produce desired outcomes. In other words, because the organizational culture influences how the firm conducts its business and helps regulate and control employees' behavior, it can be a source of competitive advantage. Given its importance, it may be that a vibrant organizational culture is the most valuable competitive differentiator for business organizations. Thus, shaping the context within which the firm formulates and implements its strategies-that is, shaping the organizational culture-is an essential strategic leadership action.

In a classic study of large U. S. corporations such as DuPont, General Motors, Chandler concluded that structure follows strategy—that is, changes in corporate strategy lead to changes in organizational structure. 钱德勒在一项经典研究中总结说,美国大型公司如杜邦、通用汽车都采取组织结构遵循战略的方式,即公司在战略上的变化导致组织结构的变化。

7.9　Establishing Balanced Organizational Controls

Organizational controls are basic to a capitalistic system and have long been viewed as an import part of strategy implementation processes. Controls are necessary to help ensure that firms achieve their desired outcomes. Defined as the "formal, information-based... procedures used by managers to maintain or alter patterns in organizational activities," controls help strategic leaders build credibility, demonstrate the value of strategies to the firm's stakeholders, and promote and support strategic change. Most critically, controls provide the parameters for implementing strategic change. Most critically, controls provide the parameters for implementing strategies as well as the corrective actions to be taken when implementation-related adjustments are required. For example, Alibaba exercised control to identify and eliminate the fraud. Furthermore, it developed additional controls to prevent such actions from occurring again.

Successful firms tend to follow a pattern of structural development, called stages of corporate development, as they grow and expand. 随着公司不断成长壮大,成功企业往往遵循某种形式的结构性发展,称为企业的发展阶段。

Stage 1：simple structure 第 1 阶段：简单结构

Stage 1 is completely centralized in the entrepreneur, who founds the company to promote an idea(product or service). 第一阶段完全集中于创立公司来推广某个想法(产品或服务)的企业家。

Stage 2：functional structure 第 2 阶段：职能结构

Stage 2 is the point when the entrepreneur is replaced by a team of managers who have functional specializations. 在第二阶段,企业家往往被具有专业职能的管理者团队取代。

Stage 3：divisional structure 第 3 阶段：分部结构

Stage 3 is typified by the corporation's managing diverse product lines in numerous industries; it decentralizes the decision-making authority. 第三阶段的典型特征是公司在许多不同的行业管理着多元化的产品线,需要分散决策权。

Stage 4：beyond SBUS 第四阶段：超越战略业务单位结构

The matrix and the network are two possible candidates for a fourth stage in corporate development-a stage that not only emphasizes horizontal over vertical connections between people and groups, but also organizes work around temporary projects in which sophisticated information systems support collaborative activities. 矩阵结构和网络结构是第四阶段公司进一步发展的两种可能候选阶段——不仅强调人与群体之间的水平和垂直连接,还围绕借助先进的信息系统支持协作活动的临时项目来组织工作。

The organizational life cycle describes how organizational grow, develop, and eventually decline. It is the organizational equivalent of the product life cycle in marketing. The stages of the organizational life cycle are Birth(stage1), Growth(stage2), Maturity(stage 3), Decline(stage 4), and Death(stage 5). 组织生命周期描述了组织如何成长、发展并最终衰退的过程,等同于营销中的产品生命周期。组织生命周期的各个阶段包括出生期(第一阶段)、成长期(第二阶段)、成熟期(第三阶段)、衰退期(第四阶段)和死亡(第五阶段)。

7.10　Innovation

Peter Drucker argued that "innovation is the specific function of entrepreneurship, whether in an existing business, a public service institution, or a new venture started by a lone individual." Moreover, Drucker suggested that innovation is "the means by which the entrepreneur either creates new wealth-producing resources or endows existing resources with enhanced potential for creating wealth." Thus, entrepreneurship and the innovation resulting from it are critically important for all firms. The realities of competition in the competitive landscape of the twenty-first century suggest that to be market leaders, companies must regularly develop innovative products desired by customers. This means that innovation should be an intrinsic part of virtually all of a firm's activities.

Innovation is a key outcome firms seek through entrepreneurship and is often the source of competitive success, especially in turbulent, highly competitive environments. For example, research results show that firms competing in global industries that invest more in innovation also achieve the highest returns. In fact, investors often react positively to the introduction of a new product, thereby increasing the price of a firm's stock. Furthermore, "innovation may be required to maintain or achieve competitive parity, much less a competitive advantage in many global markets". Investing in the development of new technologies can increase the performance of firms that operate in different but related product markets. In this way, the innovations can be used in multiple markets, and return on the investments is eared more quickly.

In his classic work, Schumpeter argued that firms engage in three types of innovative activities. Invention is the act of creating a commercial product from an invention. Invention begins after an invention is chosen for development. Thus, an invention brings something new into being, while an innovation brings something new into use. Accordingly, technical criteria are used to determine the success of an invention, whereas commercial criteria are used to determine the success of an innovation. Finally, imitation is the adoption of a similar innovation by different firms. Imitation usually leads to product or process standardization, and products based on imitation often are offered at lower prices, but without as many features. Entrepreneurship is critical to innovative activity in that it acts as the linchpin between invention and innovation.

Briefly, Six sigma is an analytical method for achieving near-perfect results on a production line. 简单来说，六西格玛是实现生产线近乎完美结果的分析方法。

Although the emphasis is on reducing product variance in order to boost quality and efficiency, it is increasingly being applied to accounts receivable, sales, and R&D. 尽管六西格玛的重点是减少产品差异，提高质量和效率，但是现在越来越的应用到应收账款、销售和研发中。

The process of six sigma encompasses five steps：六西格玛的过程包括五个步骤：

Define a process where results are poor than average. 定义一个比平均值差的过程

Measure the process to determine exact current performance. 测量过程以确定精确的目前的性能

Analyze the information to pinpoint where things are going wrong. 分析信息，找出哪里出错

Improve the process and eliminate the error. 完善过程，消除误差

Control the process to prevent future defects from occurring. 从目前发生的事情中防止未来发生错误的过程控制

In the United States in particular, innovation is the most critical of the three types of innovative activities. Many companies are able to create ideas that lead to inventions, but commercializing those inventions has, at times, proved difficult. This difficulty is suggested by the fact that approximately 80 percent of R&D occurs in large firms, but these same firms produce fewer than 50 percent of the patents. Patents are a strategic asset and the ability to regularly produce them can be an important source of competitive advantage, especially when a firm intends to commercialize the invention and when the firm competes in a knowledge-intensive industry.

7.11　Autonomous Strategic Behavior

Autonomous strategic behavior is a bottom-up process in which product champions pursue new ideas, often through a political process, by means of which they develop and coordinate the commercialization of a new good or service until it achieves success in the marketplace. A product champion is an organizational member with an entrepreneurial vision of a new good or service who seeks to create support for its commercialization. Product champions play critical roles in moving innovations forward. Indeed, in many corporations, "Champions are widely acknowledged as pivotal to innovation speed and success." Champions are vital to sell the ideas to others in the organization so that the innovations will be commercialized. Commonly, product champions use their social capital to develop informal networks within the firm. As progress is made, these networks become more formal as a means of pushing an innovation to the point of successful commercialization. Internal innovations springing from autonomous strategic behavior frequently differ from the firm's current strategy, taking it into new markets and perhaps new ways of creating value for customers and other stakeholders.

Autonomous strategic behavior is based on a firm's wellspring of knowledge and resources that are the sources of the firm's innovation. Thus, a firm's technological capabilities and competencies are the basis for new products and processes. As described in the Strategic Focus, 3M uses autonomous strategic behavior extensively to identify new technologies and products that can better serve its customers. Similarly, the iPod likely resulted from autonomous strategic behavior at Apple, thought the development of the iPhone was more the result of induced strategic behavior discussed in the next section.

Changing the concept of corporate-level strategy through autonomous strategic behavior results when a product is championed within strategic and structural contexts (see Figure 7.5). Such a transformation occurred with the development of the iPod and introduction of iTunes at Apple. The strategic context is the process used to arrive at strategic decisions. The best firms keep changing their strategic context and strategies because of the continuous changes in the current competitive landscape. Thus, some believe that the most competitively successful firms reinvent their industry or develop a completely new one across time as they compete with current and future rivals.

To be effective, an autonomous process for developing new products requires that new knowledge be continuously diffused throughout the firm. In particular, the diffusion of tacit knowledge is important for development of more effective new products. Interestingly, some of the processes important for the promotion of autonomous new product development behavior vary by the environment and country in which a firm operates. For example, the Japanese culture is high on uncertainty avoidance. As such, research has found that Japanese firms are more likely to engage in autonomous behaviors under conditions of low uncertainty because they prefer stability.

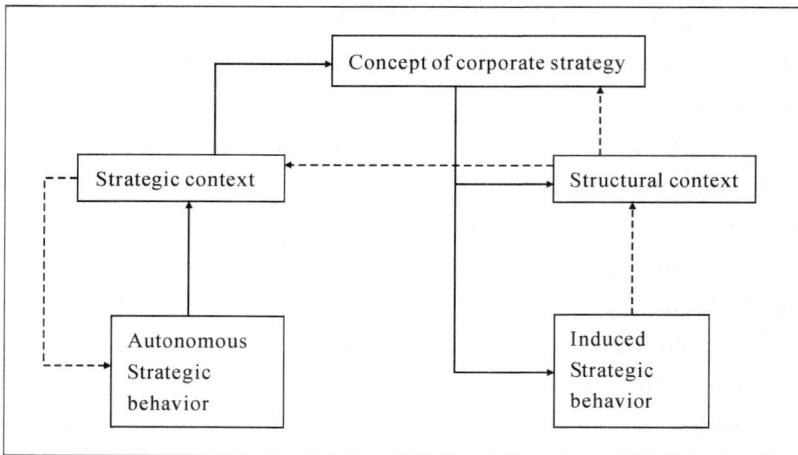

Figure 7. 5　Model of Internal Corporate Venturing

7. 12　Why the World Economy Is Globalizing

Previously closed national economies are opening up their markets to foreign companies. Importance of geographic distance is shrinking due to the Internet Growth-minded companies are racing to stake out positions in the markets of more and more countries.

What is the Motivation for Competing Internationally? Gain access to new customers. Capitalize on resource strengths and competencies Help achieve lower costs. Spread business risk across wider market base Obtain access to valuable natural resources.

International vs. Global Competition International or Multinational Competitor

132

Company operates in a few foreign countries, with modest ambitions to expand further Global Competitor Company markets products in 50 to 100 countries and is expanding operations into additional country markets annually.

Cross-Country Differences in Cultural, Demographic, and Market Conditions. Cultures and life styles differ among countries. Differences in market demographics. Variations in manufacturing and distribution costs Fluctuating exchange rates Differences in host government trade policies.

How Markets Differ from Country to Country Consumer tastes and preferences. Consumer buying habits Market size and growth potential Distribution channels Driving forces Competitive pressures One of the biggest concerns of companies competing in foreign markets is whether to customize their product offerings in each different country market to match the tastes and preferences of local buyers or whether to offer a mostly standardized product worldwide.

Potential Locational Advantages Stemming from Cost Variations Among Countries Manufacturing costs vary based on Wage rates Worker productivity. Natural resource availability Inflation rates Energy costs Tax rates Quality of a country's business environment. Clustering of suppliers, trade associations, and makers of complementary products.

Differences in Host Government Trade Policies Local content requirements Import tariffs or quotas Restrictions on exports Regulations regarding prices of imports Other regulations. Technical standards Product certification Prior approval of capital spending projects. Withdrawal of funds from country Minority ownership by local citizens.

The Role of Strategy: Actions managers take to attain the goals of the firm. Need to identify and take action that lowers the cost of value creation and/or differentiates the firm's product through superior design, quality, service, or functionality.

7. 13   Internet Strategies for Traditional Businesses

Use Internet technology to communicate and collaborate closely with suppliers and distribution channel allies Reengineer company and industry value chains to revamp how certain activities are performed and to eliminate or by pass others Make greater use of build-to-order manufacturing and assembly. Build systems to pick and pack products that are shipped individually.

Internet Strategies for Traditional Businesses (continued) Use Internet to give both existing and potential customers another choice of how to interact with the company Adopt Internet as an integral distribution channel for accessing new buyers and geographic markets Gather real-time data on customer tastes and buying habits, doing real-time market research, and use results to respond more precisely to customer needs and wants.

Key Success Factors: Competing in the E-Commerce Environment Employ an innovative business model Develop capability to quickly adjust business model and strategy to respond to changing conditions. Focus on a limited number of competencies and perform a relatively specialized number of value chain activities Stay on the cutting edge of technology Use innovative marketing techniques that are efficient in reaching the targeted audience and effective in stimulating purchases Engineer an electronic value chain that enables differention or lower costs or better value for the money.

7.14 Tailoring Strategy to Fit Specific Industry Situations

Features of an Emerging Industry New and unproven market Proprietary technology Low entry barriers Experience curve effects may permit cost reductions as volume builds Buyers are first-time users. Marketing involves inducing initial purchase and overcoming customer concerns Possible difficulties in securing raw materials Firms struggle to fund R&D, operations and build resource capabilities for rapid growth.

Strategy Options for Competing in Emerging Industries Win early race for industry leadership by employing a bold, creative strategy Push hard to perfect technology, improve product quality, and develop attractive performance features Move quickly when technological uncertainty clears and a dominant technology emerges. Form strategic alliances with Key suppliers Companies having related technological expertise.

Strategy Options for Competing in Emerging Industries(continued) Capture potential first-mover advantages Pursue New customers and user applications Entry into new geographical areas Focus advertising emphasis on. Increasing frequency of use Creating brand loyalty Use price cuts to attract price-sensitive buyers.

Features of High Velocity Markets Rapid-fire technological change Short product life-cycles Rapidly evolving customer expectations Frequent launches of new competitive moves Entry of important new rivals.

Meeting the Challenge of High-Velocity Change Source: Adapted from Shona L. Brown and Kathleen M. Eisenhardt, Competing on the Edge: Strategy as Structured Chaos. Strategic Posture Actions Strategy Reacting to Change? Introduce better products in response to new offerings of rivals? Respond to unexpected changes in buyer needs and preferences? Adjust to new government policies? React and respond as needed? Analyze prospects for market globalization? Research buyer needs, preferences, and expectations? Monitor new technological developments to predict future? Plan ahead for future changes? Add/adapt competitive capabilities? Improve product line? Strengthen distribution Leading Change? Pioneer new and better technologies? Introduce innovative products that open new markets and spur creation of whole new industries? Seek to set industry standards? Seize the offensive? Be the agent of industry change? Influence rules of the game? Force rivals to follow.

Strategy Options for Competingin High Velocity Markets Invest aggressively in R&D Develop quick response capabilities Shift resources Adapt competencies Create new competitive capabilities Speed new products to market Use strategic partnerships to develop specialized expertise and capabilities. Initiate fresh actions every few months Keep products/services fresh and exciting.

Keys to Success in Competing in High Velocity Markets Cutting-edge expertise Speed in responding to new developments Collaboration with others Agility Innovativeness Opportunism Resource flexibility First-to-market capabilities.

Characteristics of Industry Maturity Slowing demand breeds stiffer competition More sophisticated buyers demand bargains Greater emphasis on cost and service "Topping out" problem in adding production capacity Product innovation and new end uses harder to come by International competition increases Industry profitability falls Mergers and acquisitions reduce the number of industry rivals.

Strategy Options for Competing in a Mature Industry Prune marginal products and models Emphasize innovation in the value chain Strong focus on cost reduction Increase sales to present customers Purchase rivals at bargain prices Expand internationally Build new, more flexible competitive capabilities.

Strategic Pitfalls in a Maturing Industry Employing a ho-hum strategy with no distinctive features thus leaving firm "stuck in the middle" Concentrating on short-term profits rather than strengthening long-term competitiveness. Being slow to adapt competencies to changing customer expectations. Being slow to respond to price-cutting. Having too much excess capacity Overspending on marketing Failing to pursue cost reductions aggressively.

Stagnant or Declining Industries: The Standout Features Demand grows more slowly than economy as whole(or even declines) Competitive pressures intensify—rivals battle for market share To grow and prosper, firm must take market share from rivals Industry consolidates to a smaller number of key players via mergers and acquisitions.

Strategy Options for Competing in a Stagnant or Declining Industry Pursue focus strategy aimed at fastest growing market segments Stress differentiation based on quality improvement or product innovation Work diligently to drive costs down Cut marginal activities from value chain Use outsourcing Redesign internal processes to exploit e-commerce Consolidate under-utilized production facilities. Add more distribution channels Close low-volume, high-cost distribution outlets Prune marginal products.

Competing in a Stagnant Industry: The Strategic Mistakes Getting embroiled in a profitless battle for market share with stubborn rivals Diverting resources out of the business too quickly Being overly optimistic about industry's future(believing things will get better).

Competitive Features of Fragmented Industries Absence of market leaders with large market shares Buyer demand is so diverse and geographically scattered that many firms are required to satisfy buyer needs Low entry barriers Absence of scale economies Buyers require small amounts of customized or made-to-order products Market for industry's product/service may be globalizing, thus putting many companies across the world in same market arena Exploding technologies force firms to specialize just to keep up in their area of expertise Industry is young and crowded with aspiring contenders, with no firm having yet developed recognition to command a large market share.

Examples of Fragmented Industries Book publishing Landscaping and plant nurseries Auto repair Restaurant industry Public accounting Women's dresses Meat packing Paperboard boxes Hotels and motels Furniture.

Competing in a Fragmented Industry: The Strategy Options Construct and operate "formula" facilities Become a low-cost operator Specialize by product type Specialize by customer type Focus on limited geographic area.

Strategies for Sustaining Rapid Growth Companies desirous of growing revenues and earnings rapidly year-after-year have to have a portfolio of strategies. Horizon 1:Strategic initiatives to fortify and extend their position in existing businesses. Horizon 2:Strategic initiatives to leverage existing resources and capabilities by entering new businesses with promising growth potential. Horizon 3:Strategic initiatives to plant new seeds for venturing into businesses that are just emerging or do not even exist yet.

Figure 8. 2: Three Strategy Horizons for Sustaining Rapid Growth Portfolio of Strategy Initiatives? "Short-jump" initiatives to fortify and extend current businesses? Immediate gains in revenues and profits? "Medium-jump" initiatives to leverage existing resources and capabilities to pursue growth in new businesses? Moderate revenue and profit gains now, but foundation laid for sizable gains over next 2-5 years? "Long-jump" initiatives to sow the seeds for growth in businesses of the future? Minimal revenue gains now and likely losses, but potential for significant contributions to revenues and profits in 5-10 years Source:Adapted from Eric D. Beinhocker, "Robust Adaptive Time Strategies, "Sloan Management Review 40, No. 3(Spring 1999), p. 101.

Risks of Pursuing Multiple Strategy Horizons Firm should not pursue all options to avoid stretching itself too thin Pursuit of medium-and long-jump initiatives may cause firm to stray too far from its core competencies. Competitive advantage may be difficult to achieve in medium-and long-jump businesses that do not mesh well with firm's present resource strengths. Payoffs of long-jump initiatives may prove elusive.

Characteristics of Industry Leaders Stronger-than-average to powerful position Well-known reputation Proven strategies Strategic concern—How to sustain dominant leadership position Strategy Options: Industry Leaders Stay-on-the-offensive strategy Fortify-and-defend strategy Muscle-flexing strategy.

Stay-on-the-Offensive Strategies Be a first-mover, leading industry change best defense is a good offense Relentlessly pursue continuous improvement and innovation Force rivals to scramble to keep up Launch initiatives to keep rivals off balance Grow faster than industry, taking market share from rivals.

Fortify-and-Defend Strategy: Objectives Make it harder for new firms to enter and for challengers to gain ground Hold onto present market share Strengthen current market position Protect competitive advantage.

Fortify-and-Defend: Strategic Options Increase advertising and R&D Provide higher levels of customer service Introduce more brands to match attributes of rivals Add personalized services to boost buyer loyalty Keep prices reasonable and quality attractive. Build new capacity ahead of market demand Invest enough to remain cost competitive Patent feasible alternative technologies Sign exclusive contracts with best suppliers and distributors.

Muscle-Flexing Strategy: Objectives Play competitive hardball with smaller rivals that threaten leader's position Signal smaller rivals that moves to cut into leader's business will be hard fought Convince rivals they are better off playing "follow-the-leader" or else attacking each other rather the industry leader.

Muscle-Flexing: Strategic Options Be quick to meet price cuts of rivals Counter with large-scale promotional campaigns if rivals boost advertising Offer better deals to rivals' major customers Dissuade distributors from carrying rivals' products Provide sales persons with documentation about weaknesses of competing products. Make attractive offers to key executives of rivals Use arm-twisting tactics to pressure present customers not to use rivals' products.

Types of Runner-up Firms Market challengers Use offensive strategies to gain market share Focusers Concentrate on serving a limited portion of market Perennial runners-up Lack competitive strength to do more than continue in trailing position I'm trying!

Obstacles Runner-Up Firms Must Overcome When big size is a competitive asset, firms with low market share face obstacles Less access to economies of scale Difficulty in gaining customer recognition Inability to afford mass media advertising Difficulty in funding capital requirements.

Strategic Options for Runner-Up Firms When big size provides larger rivals with a cost advantage, runner-up firms have two options Build market share Lower costs and prices to grow sales or Out-differentiate rivals in ways to grow sales Withdraw from market.

Competitive Strategies for Runner-Up Firms: Building Market Share Strategic options for building market share to overcome cost advantage of larger rivals. Use lower prices to win customers from weak, higher-cost rivals Merge or acquire rivals to achieve size needed to capture greater scale economies. Invest in new cost-saving facilities and equipment, perhaps relocating operations to countries where costs are lower Pursue technological innovations or radical value chain revamping to achieve cost.

Strategic Options for Runner-Up Firms Not Disadvantaged By Smaller Size Where big size does not yield a cost advantage, runner-up firms have seven options. Offensive strategies to build market share. Growth-via-acquisition strategy. Vacant niche strategy. Specialist strategy. Superior product strategy. Distinctive image strategy. Content follower strategy.

Strategies for Runner-Up Firms Not Disadvantaged By Smaller Size Best "mover-and-shaker" offensives Pioneer a leapfrog technological breakthrough Get new/better products into market ahead of rivals and build reputation for product leadership Be more agile and innovative in adapting to evolving market conditions and customer needs Forge attractive strategic alliances with key distributors and/or marketers of similar products Find innovative ways to dramatically drive down costs to win customers from higher-cost rivals Craft an attractive differentiation strategy.

Rule of Offensive Strategy Runner-up firms should avoid attacking a leader head-on with an imitative strategy, regardless of resources and staying power an underdog may have!

Frequently used strategy of ambitious runner-up firms To succeed, top managers must have skills to Assimilate operations of acquired firms, eliminating duplication and overlap, Generate efficiency and cost savings, and Structure combined resources to create stronger competitive capabilities Growth-via-Acquisition Strategies for Runner-Up Firms.

Vacant Niche Strategies for Runner-Up Firms Focus strategy concentrated on end-use applications market leaders have neglected Characteristics of an ideal vacant niche Sufficient size to be profitable Growth potential Well-suited to a firm's capabilities Hard for leaders to serve.

Specialist Strategy for Runner-Up Firms Strategy concentrated on being a leader based on Specific technology Product uniqueness. Expertise in Special-purpose products Specialized know-how Delivering distinctive customer services.

Superior Product Strategy for Runner-Up Firms Differentiation-based focused strategy based on Superior product quality or Unique product attributes Approaches Fine craftsmanship Prestige quality Frequent product innovation Close contact with customers to gain in put for better quality product.

Distinctive Image Strategy for Runner-Up Firms Strategy concentrated on ways to stand out from rivals Approaches Reputation for charging lowest price Prestige quality at a good price Superior customer service Unique product attributes New product introductionsUnusually creative advertising.

Content Follower Strategy for Runner-Up Firms Strategy involves avoiding Trend-setting moves and Aggressive moves to steal customers from leaders Approaches Do not provoke competitive retaliation React and respond Defense rather than offense Keep same price as leaders Attempt to maintain market position.

Weak Businesses：Strategic Options Launch a strategic offensive (if resources permit) Play aggressive defense (to the extent that resources permit) Pursue immediate abandonment Adopt an end-game strategy.

The implementation of new strategies and policies often calls for new human resource management priorities and a different utilization of personnel. This may mean hiring new people with new skills, firing people with inappropriate or substandard skills, and/or training existing employees to learn new skills. 新战略和政策的实施往往需要新的人力资源管理的优先顺序和人员的不同利用。这可能意味着雇用掌握新技术的新员工,解雇技能不合适或不合格的员工,以及/或者培训现有员工学习新技能。

If growth strategies are to be implemented, new people may need to be hired and trained. 如果增长战略要实施,新人可能需要雇用和培训。

Training and development is one way to implement a company's corporate or business strategy. 培训和发展是实施一个企业的公司战略或业务战略的一种方式。

Training is also important when implementing a retrenchment strategy. 培训对于实施紧缩战略来说也是重要的。

Achieving a Turnaround：The Strategic Options Sell off assets to generate cash and/or reduce debt Revise existing strategy Launch efforts to boost revenues Cut costs Combination of efforts.

Liquidation Strategy Wisest strategic option in certain situations Lack of resources. Dim profit prospects May serve stockholder interests better than bankruptcy Unpleasant strategic option Hardship of job eliminations Effects of closing on local community.

What is an End-Game Strategy? Steers middle course between status quo and exiting quickly Involves gradually sacrificing market position in return for bigger near-term cash flow /profit Objectives Short-term-Generate largest feasible cash flow Long-term-Exit market.

Types of End-Game Options Reduce operating budget to rock-bottom Hold reinvestment to minimum. Emphasize stringent internal cost controls Place little priority on new capital investments Raise price gradually Trim promotional expenses Reduce quality in non-visible ways Curtail non-essential customer services Shave equipment maintenance.

When Should an End-Game Strategy be Considered? Industry's long-term prospects are unattractive Building up business would be too costly Market share is increasingly costly to maintain. Reduced levels of competitive effort will not trigger immediate fall-off in sales Firm can re-deploy freed-up resources in higher opportunity areas. Business is not a major component of diversified firm's portfolio of businesses Business does not contribute other desired features to overall business portfolio.

Commandments for Crafting Successful Business Strategies. Always put top priority on crafting and executing strategic moves that enhance a firm's competitive position for the long-term and that serve to establish it as an industry leader. Be prompt in adapting and responding to changing market conditions, unmet customer needs and buyer wishes for something better, emerging technological alternatives, and new initiatives of rivals. Responding late or with too little often puts a firm in the precarious position of playing catch-up.

A company can identify and prepare its people for important positions in several ways. One approach is to establish a sound performance appraisal system, which not only evaluates a person's performance, but also identifies promotion potential. 公司可以采取以下多种方式识别并准备身居重要职位的人员。一种方法是建立健全考核体系,不仅评价一个人的表现,还识别其晋升的潜力。

Many large organizations are using assessment centers, a method of evaluating a person's suitability for an advanced position. 许多大型组织使用评价中心作为评估员工是否适合未来更高层级职位的方法。

Downsizing refers to the planned elimination of positions or jobs. Companies commonly use this program to implement retrenchment strategies. 机构精简是指按照计划精简职位或工作。公司通常使用此方案实施紧缩战略。

Commandments for Crafting Successful Business Strategies. Invest in creating a sustainable competitive advantage, for it is a most dependable contributor to above-average profitability. Avoid strategies capable of succeeding only in the best of circumstances. Don't underestimate the reactions and the commitment of rival firms.

Commandments for Crafting Successful Business Strategies. Consider that attacking competitive weakness is usually more profitable than attacking competitive strength. Be judicious in cutting prices without an established cost advantage. Employ bold strategic moves in pursuing differentiation strategies so as to open up very meaningful gaps in quality or service or advertising or other product attributes.

Commandments for Crafting Successful Business Strategies. Endeavor not to get "stuck back in the pack" with no coherent long-term strategy or distinctive competitive position, and little prospect of climbing into the ranks of the industry leaders. Be aware that aggressive strategic moves to wrest crucial market share away from rivals often provoke aggressive retaliation in the form of a marketing "arms race" and/or price wars.

Implementation also involves leading: motivating people to use their abilities and skills most effectively and efficiently to achieve organizational objectives. 战略实施还涉及领导:激励员工最有效地利用自身能力和技能,高效地实现组织目标。

Leading may take the form of management leadership, communicated norms of behavior from the corporate culture, or agreements among workers in autonomous work groups. It may also be accomplished more formally through action planning or through programs such as Management by Objectives (MBO) and Total Quality Management (TQM). 领导需要领导的管理形式,沟通行为规范或与工作组的工人协调一致的企业文化。它还可能被更多的正式通过的行动计划,比如目标管理(MBO)或全面质量管理(TQM)等方式实现。

Top management responsibilities, especially those of CEO, involve getting things accomplished through and with others in order to meet the corporate objectives. 高层管理者尤其是首席执行官的责任,涉及借助他人并与他人一起实现企业目标的所有事情。

Executive leadership is the directing of activities toward the accomplishment of corporate objectives. Executive leadership is important because it sets the tone for the entire corporation. 行政领导是指导实现企业目标的各种活动。行政领导的重要性在于规定了整个公司的基调。

The CEO are able to command respect and influence strategy formulation and implementation because they tend to have three key characteristics: 首席执行官可能影响公司战略的制定和实施,他们通常有三种特征:

The CEO articulates a strategic vision for the corporation. 首席执行官拟就了公司的战略愿景。

The CEO presents a role for others to identify with and to follow. 首席执行官扮演着他人认同和遵循的角色。

The CEO not only communicates high performance standards, but also shows confidence in the followers' abilities to meet these standards. 首席执行官不仅传达高绩效标准,还给予下属有能力实现这些标准的信心。

Because an organization's culture can exert a powerful influence on the behavior of all employees, it can strongly affect a company's ability to shift its strategic direction. 因为一个组织的文化能够明显作用于所有雇员的行为,它能强烈影响一个公司转移战略方向的能力。

7.15 How Broadly a Company Should Diversify

Two questions should guide unrelated diversification efforts. What is the least diversification it will take to achieve acceptable growth and profitability? What is the most diversification that can be managed, given its added complexity? Need to strike a balance between too few different businesses and too many different businesses!

How Many Unrelated Businesses Can a Company Diversify Into? With unrelated diversification, corporate managers have to be shrewd enough to Discern good acquisitions from bad ones Select capable managers to run many different businesses. Judge soundness of strategic proposals of business-unit managers Know what to do if a subsidiary stumbles.

Diversification and Shareholder Value Related Diversification. A strategy-driven approach to creating shareholder value Unrelated Diversification. A finance-driven approach to creating shareholder value.

Combination Related-Unrelated Diversification Strategies Dominant-business firms One major core business accounting for 50-80 percent of revenues, with several small related or unrelated businesses accounting for remainder. Narrowly diversified firms Diversification includes a few (2-5) related or unrelated businesses Broadly diversified firms Diversification includes a wide ranging collection of either related or unrelated businesses or a mixture Multi-business firms Diversification portfolio includes several unrelated groups of related businesses.

When implementing a new strategy, management should consider the following questions regarding the corporation's strategy-culture compatibility-the fit between the new strategy and the existing culture：当实施一个新战略的时候,管理层应该针对公司的战略与文化的兼容性—新战略与现有文化之间的匹配,考虑以下几个问题：

Is the planned strategy compatible with the company's current culture? 公司计划的战略与目前的文化兼容吗？

Can the culture be easily modified to make it more compatible with the new strategy? 文化可以轻松改变以适应新的战略吗？

Is management willing and able to make major organizational changes and accept probable delays and a likely increase in costs? 管理层愿意并且能够做出重大组织变革,并接受可能的拖延以及成本增加吗？

Is management still committed to implementing the strategy? 管理层仍然致力于实施这一战略吗？

Strategies for Entering New Businesses Acquire existing company Internal start-up Joint venture/strategic partnerships. Acquisition of an Existing Company Most popular approach to diversification Advantages Quicker entry into target market Easier to hurdle certain entry barriers Technological in experience Gaining access to reliable suppliers. Being of a size to match rivals in terms of efficiency and costs Getting adequate distribution access.

Internal Startup More attractive when Ample time exists to create anew business from ground up Incumbents slow in responding to new entry Less expensive than buying an existing firm Company already has most of needed skills Additional capacity will not adversely impact supply-demand balance in industry New start-up does not have to go head-to-head against powerful rivals.

Joint Ventures and Strategic Partnerships Good way to diversify when Uneconomical or risky to go it alone Pooling competencies of two partners provides more competitive strength. Foreign partners are needed to surmount Import quotas Tariffs Nationalistic political interests Cultural road blocks Lack of knowledge about markets of particular countries.

Drawbacks of Joint Ventures Raises questions Which partner will do what Who has effective control Potential conflicts Control over strategy and long-term direction How operations will be conducted Control over cash flows and profits Personalities and cultures of partners.

Strategy Options for a Company Already Diversified Make new acquisitions and/or enter into additional strategic partnerships Divest some of the company's existing businesses Restructure the company's portfolio of businesses Become a multinational, multi-industry enterprise Strategy Options for a Diversified Company.

Strategies to Broaden a Diversified Company's Business Base Conditions making this approach attractive Slow grow in current business Eminently transferable resources and capabilities to other related businesses. Unexpected opportunity arises to acquire an attractive company Rapidly-changing conditions in one core industry are blurring boundaries with adjoining industries Desirable conditions favor new acquisitions to complement and strengthen market position of one or more of present businesses.

Management by Objectives (MBO) is an organization-wide approach to help assure purposeful action toward desired objectives by linking organizational objectives with individual behavior. 目标管理(MBO)是在整个组织范围内,通过将个人行为与预期目标相联系,通过将个人行为与预期目标相联系,确保有目的的行动朝着既定目标前进的方法。

MBO provides an opportunity for the corporation to connect the objectives of people at each level to those at the next higher level. 目标管理可以给公司提供一个机会,即联系每一层次的人们的目标并促使这一目标能越来越高。

Divestiture Strategies Aimed at Retrenching to a Narrower Diversification Base Strategic options Retrenchment Divestiture Spin it off as independent company Sell it Leveraged buyout Retrench? Divest? Sell? LBO?

Retrenchment Strategies Objective Reduce scope of diversification to smaller number of "core" businesses Strategic options involve divesting businesses. Having little strategic fit with core businesses Too small to contribute to earnings.

Conditions That Make Retrenchment Attractive Diversification efforts have become too broad Difficulties encountered in profitably managing broad diversification Continuing losses in certain businesses Lack of funds or resources to support operating and investment needs of all businesses. Misfits cannot be completely avoided Unfavorable changes in industry attractiveness Diversification may lack compatibility of values essential to cultural fit.

Options for Accomplishing Divestiture Spin it off as independent company Involves deciding whether to retain partial ownership or forego any ownership interest. Sell it Involves finding a company which views the business as a good deal and good fit Leveraged buy out Involves selling business to the managers who have been running it for a minimal equity down payment and loaning balance of purchase price to new owners.

Corporate Restructuring and Turnaround Strategies Strategy options for a diversified firm wit-hailing subsidiaries Why consider these options? Large losses in one or more subsidiaries Large number of businesses in unattractive industries Bad economic conditions Excessive debt load Acquisitions performing worse than expected. New technologies threatening survival of one or more core businesses. Corporate Restructuring Strategy Objective Make radical changes in mix of businesses in portfolio via both Divestitures and New acquisitions.

Conditions That Make Portfolio Restructuring Attractive Long-term performance prospects are unattractive Core business units fall upon hard times. New CEO takes over and decides to redirect where company is headed "Wave of the future"technologies emerge prompting a shakeup to build position in a new industry"Unique opportunity"emerges and existing businesses must be sold to finance new acquisition. Major businesses in portfolio become unattractive Changes in markets of certain businesses proceed in such different directions, it's better to de-merge.

Corporate Turnaround Strategies Objectives Restore money-losing businesses to profitability rather than divest them Get whole firm back in the back by curing problems of ailing businesses in portfolio Most appropriate where Reasons for poor performance are short-term Ailing businesses are in attractive industries Divesting money-losers doesn't make long-term strategic sense.

7.16   Turnaround Strategies

The Options Sell or close down a portion of operations Shift to a different, and hopefully better, business-level strategy Launch new initiatives to boost revenues Pursue cost reduction Combination of efforts.

Comment:Trend in Diversification The present trend toward narrower diversification has been driven by a growing preference to gear diversification around creating strong competitive positions in a few, well-selected industries as opposed to scattering corporate investments across many industries!

Multinational Diversification Strategies Distinguishing characteristic Diversity of businesses and diversity of national markets Presents a big strategy-making challenge Strategies must be conceived and executed for each business, with as many multinational variations as appropriate.

Appeal of Multinational Diversification Strategies Offer two avenues for long-term growth in revenues and profits Enter additional businesses. Extend operations of existing businesses into additional country markets.

Opportunities to Build Competitive Advantage via Multinational Diversification Full capture of economies of scale and experience curve effects Capitalize on cross-business economies of scope Transfer competitively valuable resources from one business to another and/or from one country to another Leverage use of a competitively powerful brand name Coordinate strategic activities and initiatives across businesses and countries Use cross-business or cross-country subsidization to out-compete rivals.

Total Quality Management (TQM) is an operational philosophy that stresses commitment to customer satisfaction and continuous improvement. 全面质量管理(TQM)是一种经营理念,强调致力于客户满意和持续改进。

TQM is committed to quality and excellence and to being the best in all functions. TQM 在所有功能中是最好的,它致力于质量和卓越。

Competitive Strength of a DMNC in Global Markets Competitive advantage potential is based on Using a related diversification strategy based on Resource-sharing and resource-transfer opportunities among businesses Economies of scope and brand name benefits Managing related businesses to capture important cross-business strategic fits Using cross-market or cross-business subsidization sparingly to secure foot holds in attractive country markets.

Competitive Power of a DMNC in Global Markets a DMNC has a strategic arsenal capable of defeating both a domestic-only rival or a single-business rival by competing in Multiple businesses and Multiple country markets.

7. 17　Financial Analysis and Enterprise Strategy

Among the existing Financial statement analysis methods, the mainstream analysis methods of the balance sheet mainly focus on the analysis of individual and overall assets, such as the analysis of debt recovery status, Inventory turnover status, fixed assets turnover status, etc. , and mainly investigate the turnover status of corresponding assets; On the other hand, the analysis of current asset turnover, Current ratio, Quick ratio, total asset turnover and total Return on assets examines the ability or quality of part structure or integrity of enterprise assets. Compared to the rich analysis methods for assets, the analysis of liabilities and shareholder equity appears very simple. The common analysis methods mainly include the analysis of the enterprise's asset liability ratio and Times interest earned to investigate the financial risk of the enterprise.

在现有财务报表分析方法中,对资产负债表的主流分析方法主要集中在对资产个体与整体的分析上,如对债权回收状况的分析、存货周转状况的分析、固定资产周转状况的分析等,主要考察相应资产的周转状况;而对流动资产周转率、流动比率、速动比率以及总资产周转率和总资产报酬率的分析等,则考察了企业资产的部分结构性或整体性的能力或者质量。与对资产的分析方法比较丰富相比,对负债和股东权益的分析就显得非常简单了。比较常见的分析方法主要有对企业资产负债率以及利息保障倍数的分析,借以考察企业的财务风险。

In fact, the competitive advantage and development potential of a company not only depend on its existing resource structure and utilization status, but also closely related to its financing environment, capital structure, corporate governance and other decisive factors in the process of enterprise development. That is to say, the real driving force of enterprise development does not lie in the size and structure of assets, nor in the strategic information reflected by the resource structure, but in the dynamic mechanism that supports enterprise development and determines enterprise strategy and its direction. The driving mechanism is determined by the right side of the enterprise's balance sheet- liabilities and shareholder equity. Among the key factors determining the development prospects and direction of enterprises, the source structure of resources (i. e. the structure of liabilities and shareholder equity) is more global and decisive than the size and structure of assets.

实际上,企业的竞争优势与发展潜力不仅取决于其现有的资源结构及其运用状况,还与其融资环境、资本结构、公司治理等在企业发展过程中具有决定意义的因素关系密切。这就是说,企业发展的真正动力不在于资产的规模和结构,不在于资源结构所反映的战略信息,而在于支撑企业发展、决定企业战略及其方向的动力机制。该动力机制是由企业资产负债表的右边——负债和股东权益来决定的。在决定企业发展前景和方向的关键性因素中,相比资产的规模与结构,资源的来源结构(即负债与股东权益的结构)更具有全局性和决定性作用。

Operational resources refer to the internal and external resources obtained by an enterprise through its daily business activities, which can also be referred to as operational liabilities. Our analysis here focuses on the ways and scale of obtaining resources, rather than specific resource forms. On the liability side of the balance sheet, the main items reflecting operating resources include notes payable, accounts payable, and advances received, as well as employee compensation and taxes payable. In accounting, it reflects the debts incurred by the enterprise in settlement with upstream and downstream enterprises, users, and employees, as well as the taxes and fees payable on the subsidiary liabilities arising from this. The main body of operational resources is the introduction or utilization of commercial credit resources by enterprises. On the one hand, under the condition that the enterprise has a higher ability to obtain commercial credit resources than the average level of similar enterprises, the enterprise usually has a high ability to "eat at both ends" (one end "eats" the upstream of the enterprise, usually suppliers; the other end "eats" the downstream of the enterprise, often distributors or consumers), Enterprises have a strong ability to use funds from upstream and downstream enterprises to support their own development; On the contrary, when a company's ability to obtain commercial credit resources is lower than the average level of similar companies, it usually means that the company's ability to "eat from both sides" is weaker and its competitive position is lower. On the other hand, commercial credit resources typically have characteristics such as low comprehensive costs (which are often lower than the average cost of loans), lower comprehensive repayment pressure than the carrying amount (corresponding repayment resources for goods or services corresponding to advance payments), and solidified upstream and downstream relationships. Maximizing the utilization of resources formed by upstream and downstream relationships has become the main resource introduction strategy for enterprises with significant competitive positions. Therefore, the utilization of commercial credit resources by enterprises is not passive or natural, but proactive and strategic-enterprises often prioritize the maximum utilization of commercial credit resources as their business and financial strategies. This means that the introduction or utilization of commercial credit resources by enterprises is not only a local issue in the management of upstream and downstream relationships, but also a strategic choice issue for enterprises. Of course, the strategic choice of enterprises to utilize commercial credit resources also depends on their competitive position or competitive advantage.

　　经营性资源是指企业通过其日常经营活动的途径所获得的内外部资源,也可以称之为经营性负债。我们这里的分析,关注的是获得资源的途径和规模,而不是具体的资源形态。在资产负债表的负债方,反映经营性资源的主要项目包括应付票据、应付账款和预收款项,以及应付职工薪酬、应交税费等。在会计核算上反映的是企业与上下游企业或者用户以及雇员进行结算时所产生的债务和由此引起的附属负债应交税费。经营性资源的主体是企业对商业信用资源的引入或者利用:一方面,在企业高于同类企业获得商业信用资源平均水平能力的条件下,企业通常具有较高的"两头吃"(一头"吃"企业的上游,往往是供应商;一头"吃"企业的下游,往往是经销商或者消费者)的能力,即企业利用上下游企业的资金来支持企业自身发展的能力较强;反之,在企业低于同类企业获得商业信用资源平均水平能力的条件下,通常意味着企业"两头吃"的能力较弱,竞争地位较低。另一方面,商业信用资源通常具有综合成本低(综合成本往往低于贷款的平均成本)、综合偿还压力低于账面金额(与预收款项对应的偿还资源为商品或劳务的账面成本)以及固化上下游关系等特点,最大限度地利用与上下游关系所形成的资源就成了具有显著竞争地位的企业主要的资源引入战略。因此，企业对于商业信用资源的利用绝不是被动、自然形成的，而是积极主动的，具有战略意义的——企业往往将最大限度利用商业信用资源作为其优先选择的经营战略与财务战略。这就是说，企业对于商业信用资源的引入或者利用，不仅仅是企业上下游关系管理的局部问题，还是企业的战略选择问题。当然，企业利用商业信用资源的战略选择还取决于其竞争地位或竞争优势。

If we break away from the concepts of liquidity of liabilities and shareholder equity, and further examine the source structure of a company's liabilities and shareholder equity, we will find that the main components of a company's liabilities and shareholder equity can be divided into four categories, namely operating resources, financial resources, shareholder investment resources, and shareholder residual resources. The utilization of these resources reflects the capital introduction strategy of the enterprise.

如果跳开负债的流动性和股东权益的概念，对企业的负债与股东权益按照其来源结构做进一步考察会发现：企业负债和股东权益的主要部分可以分成四类，即经营性资源、金融性资源、股东入资资源和股东留剩资源。对这些资源的利用体现的即是企业的资本引入战略。

Financial resources generally refer to debt financing obtained by enterprises from capital markets or financial institutions, which can also be referred to as financial liabilities. The author believes that financial resources should not only mainly come from traditional financial institutions, but also have the characteristics of financial costs (i. e. interest factors). In this way, the debt caused by financial lease in Long-term liabilities should also be financial liabilities. Therefore, in the balance sheet, in addition to typical financial resource items such as short-term loans, transactional financial liabilities, non Current liability due within one year, long-term loans, bonds payable, etc. , it should also include long-term accounts payable with interest factors and interest payable on its related ancillary debts.

金融性资源一般是指企业从资本市场或者金融机构获得的债务融资，也可以称之为金融性负债。笔者认为，金融性资源除了主要来自于传统的金融机构外，还应该具有财务代价（即利息因素）的特点。这样，在长期负债中因融资租赁而引起的债务也应属于金融性负债。因此，在资产负债表上，除了典型的金融性资源项目如短期借款、交易性金融负债、一年内到期的非流动负债、长期借款、应付债券等外，还应该包括具有利息因素的长期应付款以及与其相关的附属债务应付利息等。

If we only examine the scale and structure of financial resources (financial liabilities) of enterprises, it is easy to be introduced to pay attention to the differences in capital costs caused by different source structures and the specific projects supported by enterprise expansion, without considering the strategic implications of the structure and scale of financial resources for enterprise development. This is the constraint of accounting thinking on us. In fact, there are many factors that affect a company's choice to utilize or introduce financial resources to support its development，including the financing environment，financing costs，the company's own profitability，the fund management system of the enterprise group，the overall scale of the company's liabilities，and the existing asset liability ratio. According to the pecking order financing theory（Myers，1984），in order to maximize the interests of existing shareholders，when the enterprise has strong profitability，cannot further utilize commercial credit resources，or the scale of operational liabilities cannot meet the expansion needs of the enterprise，the enterprise will actively choose to borrow or issue bonds. Under certain shareholder investment conditions，even if the operating liabilities of the enterprise tend to zero，introducing financial resources can ensure that the expansion of the enterprise can be achieved in a certain period of time. Under the system of centralized and unified management of the internal funds of the enterprise group，the enterprise borrows or issues bonds for the financing efficiency and benefits of the whole group（not the parent company's own business activities）. Although it will increase the financial expenses on the parent company's own Income statement，it may reduce the overall financing costs of the whole group and improve the financing benefits of the whole group，which makes it the first choice of many enterprise groups' financial strategies. Therefore，analyzing and examining the utilization or introduction of financial resources in enterprises can reveal the financial strategic intentions and overall strategic planning of enterprise groups.

如果仅仅考察企业金融性资源(金融性负债)的规模和结构,就会很容易被引入关注不同的来源结构所引起的资本成本的差异以及其所支持的企业扩张的具体项目上,而不会考虑金融性资源的结构和规模对企业发展的战略含义。这就是会计思维对我们的束缚。实际上,影响企业选择利用或者引入金融性资源来支撑企业发展的因素很多,包括融资环境、融资成本、企业自身盈利能力、企业集团的资金管理体制、企业负债的整体规模以及现有资产负债率等。根据优序融资理论(Myers,1984),为了实现现有股东利益最大化,在企业具有较强的赢利能力、不能进一步利用商业信用资源或者经营性负债的规模不能满足企业扩张需求的情况下,企业会主动选择借款或者发行债券。在一定的股东入资条件下,即使企业的经营性负债趋于零,引入金融性资源也可以保证企业在一定时期的扩张得以实现。企业在实行集中统一管理企业集团内部资金的体制下,为了整个集团的融资效率与效益(不是母公司自身的经营活动)而进行借款或发行债券,尽管会增加母公司自身利润表上的财务费用,但由于可能降低整个集团的整体融资成本、提高整个集团的融资效益而成为很多企业集团财务战略的首选。因此,对企业金融性资源的利用或引入状况进行分析和考察可以看出企业集团的财务战略意图和整体战略规划。

On the balance sheet，the items reflecting shareholders' investment include capital stock（Paid-in capital）and capital reserve（this article only discusses the part caused by shareholders' investment included in capital reserve），which is the driving force of enterprise development. Shareholders have a strong strategic impact on the investment of the enterprise.

在资产负债表上,反映股东入资的项目包括股本(实收资本)和资本公积(本文仅讨论资本公积中所包含的股东入资引起的部分),这是企业发展的原动力。股东对企业的入资具有极强的战略色彩。

Ownership structure, scope of shareholders, capital scale and corporate strategy. The design of different equity structures, the selection of shareholder scope, and the arrangement of capital scale are all direct reflections of the initial strategy of the establishment stage of the enterprise. It should be noted that, in addition to having a strong leading and fundamental impact on the development of enterprises, enterprise strategies also have a dynamic feature of adjustment with the changes of enterprise operating environment, competitive position, financing environment, macro policies and other factors. Nevertheless, the investment by corporate shareholders still reflects the initial strategic intention of the company during its establishment phase. On the one hand, the degree of dispersion of equity structure and the breadth of shareholder scope directly affect the manifestation of corporate control, and it is precisely the control of the enterprise that dominates its strategy (Wang Huacheng and Hu Guoliu, 2005). On the other hand, the scale of capital also directly restricts the development strategy of enterprises: the scale of capital formed by shareholder investment is closely related to the financing ability of the enterprise, which in turn restricts the strategy and implementation of the enterprise.

股权结构、股东范围、资本规模与企业战略。不同的股权结构设计、股东范围的选择以及资本规模的安排均是企业设立阶段初始战略的直接反映。需要注意的是,企业战略对企业的发展除了具有较强的引领性和根本性影响外,还具有随企业经营环境、竞争地位、融资环境以及宏观政策等因素的变化而调整的动态性特征。尽管如此,企业股东入资仍然反映了企业设立阶段的初始战略意图。一方面,股权结构的分散程度、股东范围的广泛程度直接影响了企业控制权的表现形式,而恰恰是企业的控制权主导了企业的战略(王化成和胡国柳,2005)。另一方面,资本规模也直接制约着企业的发展战略:股东入资所形成的资本规模与企业的融资能力密切相关,进而制约着企业的战略与实施。

Ownership structure, corporate governance, core human resources and corporate strategy. Generally speaking, corporate governance needs to deal with the relationship between the shareholders' meeting (or shareholders' meeting), the board of directors, and the management of the enterprise, and ensure that the company achieves sustainable and healthy development on the basis of meeting the normal interests of all stakeholders. In the process of corporate governance, shareholders exercise their voting rights at the shareholders' meeting based on their shares, forming a board of directors. The board of directors determines the strategic goals of the company and determines core human resources; The management team led by core human resources is responsible for implementing the company's strategy. The key point of all of this is that the equity structure determines the basic structure of corporate governance.

股权结构、公司治理、核心人力资源与企业战略。一般来说,公司治理要处理的是股东大会(或股东会)、董事会与企业经理层之间的关系,并确保公司在满足各利益相关者的正常利益基础上实现持续健康发展。在公司治理的过程中,股东依其持有的股份份额在股东大会行使投票权,产生董事会。董事会决定公司的战略目标并决定核心人力资源;以核心人力资源为主导的管理团队负责实施公司的战略。而这一切的关键点在于,股权结构决定了公司治理的基本架构。

Shareholders' remaining resources refer to the portion of profits realized by a company that is not distributed by shareholders but remains in the company's equity. The remaining resources of these shareholders are mainly reflected in surplus reserves and undistributed profits on the balance sheet, which are also the accumulated profits of the enterprise. The scale of the remaining resources left by shareholders depends not only on the profitability of the enterprise, but also on the dividend or distribution policy of the enterprise.

股东留剩资源是指企业实现的利润中,股东没有分配而留存在企业的权益部分。这部分股东留剩资源在资产负债表上主要表现为盈余公积和未分配利润,也是企业的累积利润。股东留剩资源的规模形成既取决于企业的盈利能力,也取决于企业的股利或者分配政策。

The strategic significance of shareholders' remaining resources for a company lies in that, under a certain profit scale, the company can change its financial structure to a certain extent (such as changing the asset liability ratio) by formulating different dividend distribution policies (such as cash dividends, stock dividends, or a combination of the two), and form support for the company's strategy, especially financing strategy: when the company is in a high debt ratio or under high investment expenditure pressure Under the condition of relatively tight cash resources, enterprises can try their best to reduce the scale of cash dividend expenditure by choosing stock dividends or a combination of stock dividends and cash dividends, so that the shareholder's equity of the enterprise can still maintain a relatively high scale after profit distribution, thereby playing a strategic support role in reducing the cash outflow of the enterprise and improving the debt financing ability of the enterprise; On the contrary, when a company has a low debt ratio or a high asset liability ratio but a low scale of financial liabilities, abundant cash flow, and low investment cash expenditure pressure, it can choose aggressive dividend distribution policies to increase the distribution scale of cash dividends.

股东留剩资源对企业的战略含义在于,在一定的盈利规模下,企业可以通过制定不同的股利分配(如现金股利、股票股利或者是二者的组合等)政策,在一定程度上改变企业的财务结构(如改变企业的资产负债率),并对企业的战略特别是融资战略形成支撑:在企业处于高负债率或投资支出压力较大、现金资源相对紧张的条件下,企业可以通过选择股票股利或者股票股利与现金股利相结合的分配方式,尽力降低现金股利支出规模,使企业的股东权益在进行利润分配后仍然维持较高规模,从而对降低企业的现金流出量、提高企业的债务融资能力起到战略支撑的作用;反之,当企业在负债率较低、或资产负债率虽高但金融性负债规模较低、现金流量充裕、投资现金支出压力不大的条件下,可以选择激进的股利分配政策,提高现金股利的分配规模。

The above analysis clearly shows that when the debt structure of enterprises and the structure of shareholders' equity are linked with the enterprise strategy, the combination of corporate debt and shareholders' equity has far-reaching strategic implications-that is, what resources are actively used by enterprises to achieve development. Obviously, the resource utilization strategies that enterprises in different stages of development and competitive positions can adopt may be significantly different. It should be noted that when analyzing a company's balance sheet from a strategic perspective, the erterprise need to focus on overall and framework strategic information mining.

上述分析清晰地表明,当把企业的负债结构与股东权益的结构与企业战略联系起来时,企业负债和股东权益的组合状况就具有了深远的战略含义——即企业主动利用什么资源来实现发展。显然,处于不同发展阶段、不同竞争地位的企业可以采用的资源利用战略可能是显著不同的。需要说明的是,站在战略的角度对企业的资产负债表进行分析,所需要专注的是整体性和框架性的战略信息挖掘。

Chapter 8   Strategic Implementation Innovation

8. 1   Putting the Innovator's DNA into Practice

Of course, there are a variety of ways to leverage your discovery skills to make a difference. Ideally, you will uncover a big, disruptive idea, initiating meaningful change in many lives. Certainly, Bezos, Jobs, Benioff, and other innovative entrepreneurs have had an immense impact on the world. Their organizations employ hundreds of thousands of people, and their products influence—and most would say improve—the lives of hundreds of millions.

No wonder many of these business innovators moved from disrupting industries to seeking an even greater impact by aiming their attention and resources (including innovator's DNA skills) at some of the toughest world challenges, such as poverty, education, and disease.

On a smaller scale but a highly similar focus, we have also worked with social innovators around the world who rely on innovator's DNA skills to create profound solutions to some of society's most difficult problems. For example, Andreas Heinecke founded a for-profit social enterprise, Dialogue in the Dark, when working as a newspaper journalist in Germany. Heinecke's boss had brought a blind coworker to his desk and asked him to teach the person how to become a journalist. Heinecke had no idea how to approach the situation, but quickly threw himself into the task of figuring out how to make it work, in part because he had less than perfect hearing. Heinecke not only helped his blind colleague to become a journalist but, in the process, used his innovator's DNA skills to found Dialogue in the Dark, which hires blind experts to take sighted visitors into a world of complete darkness for one to three hours. (Our assessment showed Heinecke as exceptional at idea networking and questioning.) Heineke observed that to better understand and appreciate blind people, you must experience the world as they do.

In the end, most of us will likely make a difference through many minor (derivative) innovations. An idea with impact might be a new process for hiring that helps your company find more talented people (such as Google's Code Jam tournament described). It might be a new approach to marketing your company's products (such as P&G's new use of bloggers and customer generated content). Or it might be building a business model based on the premise that for every pair of shoes sold, the company will give away one

153

pair, as Blake Mycoskie did when he founded TOMS Shoes (after traveling to Argentina in 2006 and seeing so many children with foot diseases because they lacked shoes).

Clearly, the process of creative discovery can be difficult, but the rewards far outstrip the challenges. Being a creator is exciting, and to author or coauthor an idea that leads to a new product, service, process, or business energizes. Being an innovator is psychologically and emotionally gratifying in a way that money simply isn't, even though the financial rewards of successful innovation can be significant. Mark Ruiz, co-founder of Micro Ventures and finalist for the Entrepreneur of the Year Philippines 2010 award, admitted the same when he told us, "even though I'm an entrepreneur, what drives me is not really the money. What really drives me is a deep sense of mission and purpose. I just see problems that are screaming for new and innovative solutions." Ruiz works nonstop to build new venture after new venture to take on these problems in his home country, the Philippines.

Say that you have one of those reality shows on TV and you drop a bunch of people in the middle of a desert island. The winner is the person who gets to the shore the quickest. Some people try to analyze where they are, which direction to go. Some of them say, "Let's climb up a tree or a rock or hillside and maybe we can see further and figure out what is the best direction to go." They will spend time planning and analyzing how to find the best direction to go. But some other people will just look around, follow their intuition, and start running in a direction.

If there are a lot of people that have been dropped on the island, I can almost guarantee that whoever starts climbing up the tree to start analyzing where he is and which direction to go will not win the competition. Why? Because there are a few other maniacs who will follow their intuition and just start running. They're much more likely to get to shore quicker. The point is: if you have a good gut feeling for which general direction to go, then you should just run as fast as you can.

In the end, innovation is an investment—in yourself, in others, and if you're a senior manager or emerging entrepreneur, in your company. Whether you're working at the top of an organization or as a technical specialist at the bottom, eBay's Whitman advises everyone "to have the courage to plant acorns before you need oak trees." Innovation is all about planting acorns (ideas) with less than complete confidence that each will grow into something meaningful. The alternative, however, is little or no growth when no acorns emerge as trees. By understanding and reinforcing the DNA of individual innovators within innovative teams and organizations, you can find ways to more successfully develop not just growth saplings but the real oak trees of future growth. As you continue your innovation journey, let your life speak2 the final line from Apple's Think Different campaign: "The people who are crazy enough to think they can change the world are the ones who do. " So just do it. Do it now!

Not surprisingly, the same innovators worked equally hard to infuse a parallel set of taken-for-granted philosophies deep into every nook and cranny of their company's culture (just as Bezos did at Amazon). They recognized that a culture is most powerful when widely shared and deeply held. So how did they do this? They knew that their own innovation example was an important first step to building a highly innovative company. They also realized it was impossible to personally lead or participate in every team and that they would have limited direct contact with most employees (especially as their companies grew). As such, they worked hard to instill a deep, companywide commitment to innovation. Not only did their companies pay attention to picking innovative people and putting innovative processes into place, they also lived by a set of key innovation philosophies.

Here's what innovative entrepreneurs and executives told us about their innovation philosophies. We heard that innovation is everyone's job. We learned that disruptive innovation is part of their company's innovation portfolio. We found out that having lots of small project teams, properly organized, is central to the way their companies took innovative ideas to market. Finally, we realized that they *do* take more risks than other companies in the pursuit of innovation, but they take actions that mitigate those risks, thereby turning them into "smart risks. " These four philosophies permeate the world's most innovative companies and are not only expressed through words but punctuated powerfully through reinforcing actions.

In rejecting the limiting belief that innovation is R&D's job alone, leaders of highly innovative companies—such as Jobs, Bezos, and Benioff—work hard to instill "innovation is everyone's, job" as a guiding organizational philosophy. When Jobs returned, to Apple after a twelve-year hiatus, he launched the "Think Different" advertising campaign. The ad paid tribute to a wide range of innovators saying, "Here's to the crazy ones. The misfits. The rebels. The trouble makers . . . the ones who see things differently. They're not fond of rules. And they have no respect for the status quo . . . they change things. They push the human race forward. " The Emmy award-winning Think Different campaign was hailed as one of the most innovative of all time, largely because it inspired people. What most people don't realize, though, is that the campaign targeted Apple employees as much as its customers. "The whole purpose of the 'Think Different' campaign ,was that people had forgotten what Apple stood for, including the employees," said Jobs. "We thought long and hard about how you tell somebody what you stand for, what your values are, and it occurred to us that if you don't know somebody very well, you can ask them, 'Who are your heroes?' You can learn a lot about people by hearing who their heroes are. So we said, 'Okay, we'll tell them who our heroes are. '" To

155

reestablish Apple's innovativeness, jobs knew that every employee needed this message: "Our heroes are innovators. We stand for innovation. If you want to work at Apple, we expect you to be an innovator who wants to change the world."

The Think Different campaign is just one of many things Jobs has done to send the message to Apple employees that innovation is their job. He once urged the original Macintosh development team to innovate by saying, "Let's make a dent in the universe. We'll make it so important that it will make a dent in the universe." More recently, he encouraged Disney employees to "dream bigger"(as the largest single shareholder of The Walt Disney Company stock, Jobs has a vested interest in Disney being innovative). These bold statements send a clear message to employees: we expect each of you to innovate.

Bold actions must follow bold statements to reinforce the message. P&G's Lafley pursued the "we innovate" philosophy when he remarked, "The P&G of five or six years ago depended on eight thousand scientists and engineers for the vast majority of innovation. The P&G we're trying to unleash today asks all hundred thousand-plus of us to be innovators." To reinforce his commitment to organization wide innovation, he actively solicited ideas from throughout the company, and if the concept showed promise, he put it into development. For example, Lafley backed a successful hair-care product line for women of color because a few African American employees explained to him that existing products didn't work well and "we can do better." P&G did better, launching a successful new line, Pantene Pro-V Relaxed & Natural. Lafley's actions set the tone for innovate philosophy to take hold. Yet, key leaders' personal actions alone are not enough. We saw that highly innovative companies, compared to typical companies, reinforce this philosophy by giving people more time and resources to innovate. Jobs knew that every employee needed this message: "Our heroes are innovators. We stand for innovation. If you want to work at Apple, we expect you to be an innovator who wants to change the world."

The Think Different campaign is just one of many things Jobs has done to send the message to Apple employees that innovation is their job. He once urged the original Macintosh development team to innovate by saying," Let's make a dent in the universe. We'll make it so important that it will make a dent in the universe."More recently, he encouraged Disney employees to "dream bigger"(as the largest single shareholder of The Walt Disney Company stock, Jobs has a vested interest in Disney being innovative). These bold statements send a clear message to employees: we expect each of you to innovate.

8. 2　Technology in services

As mentioned in the paragraph above, many services are faced with increasing demand

and customer contact. To help address the problems services are turning to technology. The National Health Service is a good example. Struggling to cope with ever-increasing demands, the introduction of new technology will now play a much larger part in delivering medical care. In addition to the out-of-hours phone service, patients will be diagnosed and monitored from afar by using telemedicine facilities and the Internet. With help and support from a health professional, the patient will play an increasing role in managing his/her own condition. Unlike the examples of the petrol pump or self-service in the supermarket, this example provides a more substantive, if controversial, illustration of the customer as part-time employee or co-producer of a service. Furthermore, the advance of technology not only enables services to transfer some of the work to customers but also allows organizations to vary according to customer type, the service received. With the advent of information technology, organizations can now amass a substantial amount of data on customers. Those seen as less worthy, less profitable or perceived to be a cause of inefficiency (e. g. elderly people blocking hospital beds) will need to serve themselves or simply go away.

Welcome to the new consumer apartheid. Those long lines and frustrating telephone trees aren't always the result of companies simply not caring about pleasing the customer anymore. Increasingly, companies have made a deliberate decision to give some people skimpy service because that's all their business is worth. Call it the dark side of the technology boom, where marketers can amass a mountain of data that gives them an almost Orwellian view of each buyer. Consumers have become commodities to pamper, squeeze, or toss away, according to Leonard Berry, marketing professor at Texas A&M University. He sees 'a decline in the level of respect given to customers and their experiences'.

The benefits of technology, for service organizations, are apparent in terms of productivity and cost savings. But what of the customer? After all, 'much of this technology delivered service is initiated and carried out by the consumer and involves no direct or indirect contact with representatives of the service provider'. This is given further meaning by the view that 'across industries, technology is dramatically altering interpersonal encounter relationships, and in some instances, eliminating them altogether'. So in this age of increasing technology-based service, what are the views of the customers? Dabholkar studied a situation in a fast-food restaurant where customers could use a computerized touch screen to order a meal (technology- based self-service). The study found that feeling in control and the potential enjoyment from using this type of delivery, were important determinants of service quality. Reliability (error free), speed of delivery and ease of use were also found to be important. In an extensive study of consumers, Howard and boys found that time-saving was seen as the biggest advantage of self-service. However, the findings also suggest that consumers still prefer the concept of human interaction rather than technological interfaces. Similar findings by Curry and Penman from the banking sector indicate that the human element in the banker/customer relationship is more influential than the technology element. Equally they draw attention to differences between banks and their employees in the use of technology.

In a review of electronic government in the public sector, Hazlett and Hill point to a 'lack of evidence to support the claim that the use of technology in service delivery results in less bureaucracy and increased quality'. They conclude that 'it is by no means certain that e-government can produce truly innovative, responsive public services, indeed it may merely exacerbate electronically, existing shortcomings'. It would appear that the overriding problem for technology in services is the willingness and ability of customers to use it. Parasuraman has developed a scale called the technology readiness index designed to classify people in terms of their tendency to embrace and use new technology. The classification is arranged hierarchically with explorers at the top and laggards at the bottom. Not surprisingly, research indicates that those most favourably disposed to technology and heavy users are well educated, with a high income and younger. Those at the other end tend to be older, of a lower education level and income. What is intriguing is how far into the future this particular way of classifying people (in relation to technology use) will persist. Even more so, if you recall an earlier reference and with some worry, those at the lower or bottom end are the same group for whom technology will be the 'option' for accessing a range of services.

Technology not only appears in service provider/customer contacts. It is also present within the service organization's work environment. The aims are largely twofold: management control and efficiency. The following examples are simply illustrative. Ocado, an upmarket online grocer and a favourite with wealthier stay-at-home shop- pers in London and the South East, has been the subject of a study34 by the trade union GMB (General and Municipal Boilermakers). It centres on the company's practices at its distribution centre at Hatfield, where its warehouse workers, in company with an estimated 10 000 in all among other named firms (Tesco, Marks & Spencer, Sainsbury, B&Q, Boots and Homebase), are compelled to strap on chunky electronic surveillance tags that direct them to pick up goods for delivery. Professor Michael Blakemore of Durham University (the report author) says satellite and radio-based computer technology is turning some warehouses into 'battery farms' and creating 'prison surveillance'. Employers argue the union is over-reacting. They insist the system is not used to spy on workers but to make things easier and improve customer service. However, Paul Kenny, the union's acting General-Secretary, says: 'The GMB is not a Luddite organization, but we will not stand idly by to see our members reduced to robots with heartbeats.'

Services and products are not created in a vacuum. They are produced by organizations that vary in size, structure and culture. Knowledge of how organizations operate, and why, assists our understanding of behaviour in organizations. This is particularly important for services, as customers are involved in varying degrees in the production and delivery process. According to one leading authority in services marketing, 'how to organize to implement the services strategy is among the most crucial of decisions'. 1 Unfortunately, the literature on organizational behaviour (as expressed in standard textbooks) is contested by some as not adequately reflecting organizational realities-realities that may obstruct any move toward determining the 'optimum organization for service'.

Initially we thought of our task as writing an alternative organizational book, one which redressed the inaccurate account of organizational behavior contained in many texts in our field. Standard textbooks in this area say surprisingly little about the character of the phenomena with which they are centrally concerned-the behavior routinely exhibited by people in organizations. What they do say suggests, as much by implication as direct assertion, that behavior in organizations is, almost conforming and dutiful. Besides shedding light on what managers really do, observational studies also provide insights on why managers often find contemporary models of management and organizational behavior discrepant with the real world and why they frequently do not follow the resulting prescriptions.

The main finding was that the focal group's estimate of how their internal customers would rate it on a number of variables, e. g. courtesy, competence, prompt service, was significantly higher than the actual ratings attributed to the focal group by its internal customers. Although the hierarchical form is giving way to more horizontal communication and coordination, it remains a defining characteristic of organizations. Relationships between top and bottom of the organization offer up as much, if not greater, potential for influencing the quality of service as do relationships across the organization. Unlike the horizontal, however, the vertical displays much greater disparities of power, status, authority and general working conditions. Some advocate inverting the levels of the traditional pyramid as a way of generating and delivering service , excellence The main finding was that the focal group's estimate of how their internal customers would rate it on a number of variables, e. g. courtesy, competence, prompt service, was significantly higher than the actual ratings attributed to the focal group by its internal customers.

Although the hierarchical form is giving way to more horizontal communication and coordination, it remains a defining characteristic of organizations. Relationships between top and bottom of the organization offer up as much, if not greater, potential for influencing the quality of service as do relationships across the organization. Unlike the horizontal, however, the vertical displays much greater disparities of power, status, authority and general working conditions. Some advocate inverting the levels of the traditional pyramid as a way of generating and delivering service excellence.

Trying to understand and explain what happens in organizations can be a frustrating experience. The concept of corporate culture is viewed as offering some hope for unravelling the mystique of organizational life. The essence of culture is reserved for the deep level of basic assumptions and beliefs shared by members of an organization that operate unconsciously and define (in a basic taken-for-granted fashion) an organization's view of itself and its environment.

The nature of the service operation can therefore pigeonhole an organization. The UK National Health Service, however, is an interesting example of the clash of two cultures— the caring professionals and the business efficiency of the administrators. The personal culture of the professionals like lawyers is giving way to more organizational disciplines. The support culture has historically been associated with the social services but any service organization or individuals within it can adopt the values of caring, responsiveness, receptivity and a sense of belonging.

Overall, what is clear is that regardless of the type of service there is a growing emphasis on areas such as organization, management, efficiency, budgets and performance measurement. In other words, 'progress' toward a 'business culture'. A major feature of the literature on corporate culture in general, and for service excellence in particular, is that perceptions and beliefs about what is important and why those things are important should be shared by employees throughout the organization. This is necessary if an organization is to be effective in its basic processes-communication, cooperation, commitment, decision making and implementation. Through recruitment and selection policies and socialization processes (training and development) organizations seek to shape employee beliefs and behavior in accordance with the prevailing values and norms of the organization. However, organizational culture can be a 'contested reality'. Not everyone or every group always shares the established view.

8.3   Services Marketing：Concepts，Strategies

The molecular model is another useful tool for expanding our understanding of the basic differences between goods and services. A molecular model is a pictorial representation of the relationship between the tangible and intangible elements of a firm's operation. One of the primary benefits obtained from developing a molecular model is that it is a management tool that offers the opportunity to visualize the firm's entire bundle of benefits that its product offers customers. Depicts two molecular models which continue our earlier discussion concerning the differences between automobile ownership (tangible dominant) and purchasing an airline ticket (intangible dominant). As previously discussed, airlines differ from automobiles in that consumers typically do not physically possess the airline. Consumers in this case purchase the core benefit of transportation all the corresponding tangible (denoted by solid-lined circles) and intangible benefits (denoted by dashed-lined circles) that are associated with flying. In contrast, a consumer who purchases an automobile primarily benefits by the ownership of a physical possession that renders a service—transportation.

The overriding benefit obtained by developing molecular models is the appreciation for the intangible and tangible elements that comprise most products. Once managers understand this broadened view of their products, they can do a much better job of understanding customer needs, servicing those needs more effectively, and differentiating their product-offering from competitors. The molecular model also demonstrates that consumers' service "knowledge" and goods "knowledge" are not obtained in the same manner. With tangible dominant products, goods "knowledge" is obtained by focusing in on the physical aspects of the product itself. In contrast, consumers evaluate intangible dominant products based on the experience that surrounds the core benefit of the product. Hence, understanding the importance and components of the service experience is critical.

Due to the intangible nature of service products, service knowledge is acquired differently than knowledge pertaining to goods. For example, consumers can sample tangible dominant products such as soft drinks and cookies prior to purchase. In contrast, a consumer cannot sample an intangible dominant product such as a haircut, a surgical procedure, or a consultant's advice prior to purchase. Hence, service knowledge is gained through the experience of receiving the actual service itself. Ultimately, when a consumer purchases a service, he or she is actually purchasing an experience!

All products, whether goods or services, deliver a bundle of benefits to the consumer. The benefit concept is the encapsulation of these tangible and intangible benefits in the consumer's mind. For a tangible dominant good such as Tide laundry detergent, for example, the core benefit concept might simply be cleaning. However for other individuals, it might also include attributes built into the product that go beyond the mere powder or liquid, such as cleanliness, whiteness, and/or motherhood (it's a widely-held belief in some cultures that the cleanliness of children's clothes is a reflection upon their mother). The determination of what the bundle of benefits comprises—the benefit concept purchased by consumers—is the heart of marketing, and it transcends all goods and services.

In contrast to goods, services deliver a bundle of benefits through the experience that is created for the consumer. For example, most consumers of Tide laundry detergent will never see the inside of the manufacturing plant where Tide is produced. Customers will most likely never interact with the factory workers who produce the detergent or with the management staff that directs the workers. In addition, customers will also generally not use Tide in the company of other consumers. In contrast, restaurant customers are physically present in the "factory" where the food is produced; these customers do interact with the workers who prepare and serve the food as well as with the management staff that runs the restaurant. Moreover, restaurant customers consume the service in the presence of other customers where they may influence one another's service experience. One particularly simple but powerful model that illustrates factors that influence the service experience is the model depicted.

Another important aspect of the customer's experience involves the contact personnel and service providers that directly interact with customer. Technically speaking, contact personnel are employees—other than the primary service provider—who briefly interact with the customer. Typical examples of contact personnel are parking attendants, receptionists, and hosts and hostesses. In contrast, service providers are the primary providers of the core service, such as a waiter or waitress, dentist, physician, or college instructor. Unlike the consumption of goods, the consumption of services often takes place where the service is produced (e. g. , dentist's office, restaurant, and hairstylist) or where the service is provided at the consumer's residence or workplace (e. g. , lawn care, house painter, janitorial service). Regardless of the service delivery location, interactions between consumers and contact personnel/service providers are commonplace.

As a result, service providers have a dramatic impact on the service experience. Another important aspect of the customer's experience involves the contact personnel and service providers that directly interact with customer. Technically speaking, contact personnel are employees—other than the primary service provider—who briefly interact with the customer. Typical examples of contact personnel are parking attendants, receptionists, and hosts and hostesses. In contrast, service providers are the primary providers of the core service, such as a waiter or waitress, dentist, physician, or college instructor. Unlike the consumption of goods, the consumption of services often takes place where the service is produced (e. g. , dentist's office, restaurant, and hairstylist) or where the service is provided at the consumer's residence or workplace (e. g. , lawn care, house painter, janitorial service). Regardless of the service delivery location, interactions between consumers and contact personnel/service providers are commonplace. As a result, service providers have a dramatic impact on the service experience.

Service personnel perform the dual functions of interacting with customers and reporting back to the internal organization. Strategically, service personnel are an important source of product differentiation. It is often challenging for a service organization to differentiate itself from other similar organizations in terms of the benefit bundle it offers or its delivery system. For example, many airlines offer similar bundles of benefits and fly the same types of aircraft from the same airports to the same destinations. Therefore, their only hope of a competitive advantage is from the service level—the way things are done. Hence, the factor that often distinguishes one airline from another is the poise and attitude of its service providers. Singapore Airlines, for example, enjoys an excellent reputation due in large part to the beauty and grace of its flight attendants. Other firms that hold a differential advantage over competitors based on personnel include the Ritz Carlton, IBM, and Disney Enterprises.

The service sector is one of the three main categories of a developed economy—the other who are industrial and agricultural. Traditionally, economies throughout the world tend o transition from an agricultural economy to an industrial economy (e. g. , manufacturing, mining, etc.) to a service economy. The United Kingdom was the first economy in the modern world to make this transition. Several other countries including the United States, Japan, Germany and France have made this transition, and many more are expected to do so at an accelerated rate. We live in interesting times! The increased rate of transformation from an agricultural to a manufacturing to a service-based economy has generally been caused by a highly competitive international marketplace. Simply stated, goods are more amenable to international trade than services, thereby, marking them more vulnerable to competitive actions.

In other words, countries that industrialized their economies first eventually come under attack by other countries that are newly making the transition from an agricultural to an industrial economy. These "newcomer" countries offer lower production costs (especially labor) which attract industry. Consequently, as industrial sectors flow from one country to the next, the countries they abandon begin to more heavily rely on the growth of their service sectors as the mainstay of their economies. This whole process repeats itself over and over again as other less developed countries enter the fray, consequently facilitating the transformation from agriculture to industrial to service-based economies—which in turn has created economic growth throughout the world.

One way of viewing the relationship between marketing and operations in a service firm is to think of it as marrying the consumers' needs with the technology and manufacturing capabilities of the firm. As is the case with most marriages, a successful union between operations and marketing will obviously involve compromises, since the consumers' needs can seldom be met completely and economically. In a service firm, establishing a balance between marketing and operations is critical! Significant aspects of the service operation are the product that creates the experience that delivers the bundle of benefits to the consumer. For example, a restaurant experience is not based solely on the quality of the food. The physical environment and interactions with contact personnel throughout the experience also greatly influence consumer perceptions of the quality of service delivered. A successful compromise between operations efficiency and marketing effectiveness is, therefore, that much more difficult to achieve. Consequently, success in services marketing demands a much greater understanding of the constraints and opportunities posed by the firm's operations. To introduce these complexities, in this chapter we first adopt the perspective of an operations manager and ask, "What would be the ideal way to run the service delivery system from an operations perspective?" The impact on marketing and the opportunities for marketing to assist in the creation of this ideal are then developed.

Obviously, such an ideal world as proposed by Thompson is virtually impossible to create, and even in goods companies the demands of purchasing the inputs and marketing's management of the outputs have to be traded off against the ideal operations demands. In goods manufacturing, this tradeoff has been accomplished through the focused factory. The focused factory focuses on a particular job; once this focus is achieved, the factory does a better job because repetition and concentration in one area allow the workforce and managers to become effective and experienced in the task required for success. The focused factory broadens Thompson's perfect-world model in that it argues that focus generates effectiveness as well as efficiency. In other words, the focused factory can meet the demands of the market better whether the demand is low cost through efficiency, high quality, or any other criterion.

8.4 Meeting Competitive Challenges and Acquiring Critical Resources

As a firm analyzes its external environment and assesses its internal resources and capabilities to meet environmental challenges, acquisitions as well as adjustments to the firm's set of businesses are often considered. Domestically, a number of U. S. firms have found that horizontal acquisitions (acquisitions of potential competitors) meet their needs to handle these environmental challenges and resource considerations. For example, Sears and Kmart merged in order to meet the competitive challenge by Wal-Mart and other large discount retailers. Likewise, consumer product firms Gillette and Procter & Gamble merged. Phone companies have partially met aggressive challenges from cable companies offering local phone service by making acquisitions in the long distance area. For instance, SBC Communications acquired AT&T's long distance business. In response to this move, Verizon and Quest have been battling over the opportunity to acquire MCI, a long distance service company. Most of these horizontal acquisitions have been directed at obtaining ore efficiency and market power. Theirs have been directed at diversifying into new areas of business where the competitive challenge is not as significant as it is in the telecommunications acquisitions.

Still other horizontal acquisitions have been undertaken in order to respond to industry overcapacity. For instance, the proposed acquisition of America West Airlines by U. S. Airways will create more critical mass for the merged airline to compete with larger legacy carriers as well as with discounters such as Southwest Airlines and Air Tran Airways. Once the merger is complete, this combination will allow these airlines to reduce some of the overcapacity in the industry. There have also been a number of cross-border acquisitions announced, especially Chinese firms seeking to obtain opportunities and especially critical resources allowing them to compete in the important U. S. market. Many of these acquisitions appear to be horizontal. In 2005, Lenovo Group, the largest personal computer manufacturer in China, acquired the PC assets of IBM and is allowed to use the IBM brand label for five years following the acquisition. Lenovo plans to introduce its own brand in association with IBM during this five-year period to build up brand equity in the U. S. market, a critical resource necessary to compete globally.

Similarly, the Haier Group, the largest manufacturer of appliances in China, proposed (but later withdrew its offer) to purchase Maytag Corp. in order to build its presence in the United States through the Maytag brand. Chinese National Offshore Oil Corporation (CNOOC), a major oil and natural gas producer in the Chinese domestic market, sought to break up a deal between Chevron and Unocal Corp. by offering a higher takeover bid for Unocal.

Although the frequency of acquisitions has slowed, their number remains high. In fact, an acquisition strategy is sometimes used because of the uncertainty in the competitive landscape. A firm may make an acquisition to increase its market power because of a competitive threat, to enter a new market because of the opportunity available in that market, or to spread the risk due to the uncertain environment. In addition, as volatility brings undesirable changes to its primary markets, a firm may acquire other companies to shift its core business into different markets. Such options may arise because of industry or regulatory changes. For instance, Clear Channel Communications built its business by buying radio stations in many geographic markets when the Telecommunications Act of 1996 changed the regulations regarding such acquisitions. However, more recently Clear Channel has been suggested to have too much market power and is now likely to split into three different businesses.

8.5　Acquisition Strategy

The strategic management process calls for an acquisition strategy to increase a firm's strategic competitiveness as well as its returns to shareholders. Thus, an acquisition strategy should be used only when the acquiring firm will be able to increase its value through ownership of an acquired firm and the use of its assets. However, evidence suggests that, at least for the acquiring firms, acquisition strategies may not always result in these desirable outcomes. Researchers have found that shareholders of acquired firms often earn above-average returns from an acquisition, while shareholders of acquiring firms are less likely to do so, typically earning returns from the transaction that are close to zero. In the latest acquisition boom between 1998 and 2000, acquiring firm shareholders experienced significant losses relative to the losses in all of the 1980s. Acquiring firm shareholders lost $0.12 on average for the acquisitions between 1998 and 2000 whereas in the 1980s shareholders lost $.016 per dollar spent. This may suggest that for large firms, it is now more difficult to create sustainable value by using an acquisition strategy to buy publicly traded companies.

In approximately two-thirds of all acquisitions, the acquiring firm's stock price falls immediately after the intended transaction is announced. This negative response is an indication of investors' skepticism about the likelihood that the acquirer will be able to achieve the synergies required to justify the premium. Although the frequency of acquisitions has slowed, their number remains high. In fact, an acquisition strategy is sometimes used because of the uncertainty in the competitive landscape. A firm may make an acquisition to increase its market power because of a competitive threat, to enter a new market because of the opportunity available in that market, or to spread the risk due to the uncertain environment. In addition, as volatility brings undesirable changes to its primary markets, a firm may acquire other companies to shift its core business into different markets. Such options may arise because of industry or regulatory changes. For instance, Clear Channel Communications built its business by buying radio stations in many geographic markets when the Telecommunications Act of 1996 changed the regulations regarding such acquisitions. However, more recently Clear Channel has been suggested to have too much market power and is now likely to split into three different businesses.

The strategic management process calls for an acquisition strategy to increase a firm's strategic competitiveness as well as its returns to shareholders. Thus, an acquisition strategy should be used only when the acquiring firm will be able to increase its value through ownership of an acquired firm and the use of its assets. However, evidence suggests that, at least for the acquiring firms, acquisition strategies may not always result in these desirable outcomes. Researchers have found that shareholders of acquired firms often earn above-average returns from an acquisition, while shareholders of acquiring firms are less likely to do so, typically earning returns from the transaction that are close to zero. In the latest acquisition boom between 1998 and 2000, acquiring firm shareholders experienced significant losses relative to the losses in all of the 1980s. Acquiring firm shareholders lost $ 0. 12 on average for the acquisitions between 1998 and 2000 whereas in the 1980s shareholders lost $. 016 per dollar spent. This may suggest that for large firms, it is now more difficult to create sustainable value by using an acquisition strategy to buy publicly traded companies. In approximately two-thirds of all acquisitions, the acquiring firm's stock price falls immediately after the intended transaction is announced. This negative response is an indication of investors' skepticism about the likelihood that the acquirer will be able to achieve the synergies required to justify the premium.

8.6 Merger, takeover, acquisition

A merger is a strategy through which two firms agree to integrate their operations on a relatively coequal basis. There are few true mergers, because one party is usually dominant in regards to market share or firm size. Daimler Chrysler AG was termed a "merger of equals" and, although Daimler-Benz was the dominant party in the automakers' transaction, Chrysler managers would not allow the business deal to be completed unless it was termed a merger. An acquisition is a strategy through which one firm buys a controlling, or 100 percent, interest in another firm with the intent of making the acquired firm a subsidiary business within its portfolio. In this case, the management of the acquired firm reports to the management of the acquiring firm. While most mergers are friendly transactions, acquisitions can be friendly or unfriendly.

A takeover is a special type of an acquisition strategy wherein the target firm does not solicit the acquiring firm's bid. The number of unsolicited takeover bids increased in the economic downturn of 2001 – 2002, a common occurrence in economic recessions; because the poorly managed firms that are undervalued relative to their assets are more easily identified. Many takeover attempts are not desired by the target firm's managers and are referred to as hostile. In a few cases, unsolicited offers may come from parties familiar and possibly friendly to the target firm. On a comparative basis, acquisitions are more common than mergers and takeovers. Accordingly, this chapter focuses on acquisitions.

A primary reason for acquisitions is to achieve greater market power. Market power exists when a firm is able to sell its goods or services above competitive levels or when the costs of its primary or support activities are below those of its competitors. Market power usually is derived from the size of the firm and its resources and capabilities to compete in the marketplace. It is also affected by the firm's share of the market. Therefore, most acquisitions that are designed to achieve greater market power entail buying a competitor, a supplier, a distributor, or a business in a highly related industry to allow the exercise of a core competence and to gain competitive advantage in the acquiring firm's primary market. One goal in achieving market power is to become a market leader. 23 In 2005, Federated Department Stores, Inc. completed an acquisition of May Department Stores Co. This represents a horizontal acquisition in the large department store retail segment of the "big box" retail store industry. Both Federated and May have been squeezed at the discount end by Wal-Mart and at the high luxury end by firms such as Neiman Marcus.

This acquisition represents Federated's hope that by increasing the size of the firm it can maintain enough efficiency to be competitive. Federated is the parent of Macy's and Bloomingdale's, while May's chains include Lord & Taylor, Marshall Field's, and Filene's. These two firms are a good match geographically and across concept (e. g. , high quality versus cost conscious), but some malls will probably lose stores because of the overlap in store concepts. This will, however, improve local market power and reduce costs for these businesses. Research in marketing suggests that performance of the merged firm increases if marketing-related issues are involved. The performance improvement of the merged firm subsequent to a horizontal acquisition is even more significant than the average potential cost savings if marketing of the combined firms improves economies of scope. To increase their market power, firms often use horizontal, vertical, and related acquisitions.

The acquisition of a company competing in the same industry as the acquiring firm is referred to as a horizontal acquisition. Horizontal acquisitions increase a firm's market power by exploiting cost-based and revenue-based synergies. Research suggests that horizontal acquisitions result in higher performance when the firms have similar characteristics. Examples of important similar characteristics strategy, managerial styles, and resource allocation patterns. Similarities in these characteristics

make the integration of the two firms proceed more smoothly. Horizontal acquisitions are often most effective when the acquiring firm integrates the acquired firm's assets with its assets, but only after evaluating and divesting excess capacity and assets that do not complement the newly combined firm's core competencies. As the acquisition of May by Federated illustrates, the merged firm will likely have to divest itself of some stores in order to reduce costs associated with the acquisition.

8.7 Vertical Acquisition

A vertical acquisition refers to a firm acquiring a supplier or distributor of one or more of its goods or services. A firm becomes vertically integrated through this type of acquisition in that it controls additional parts of the value chain. Kodak's acquisition of Creo, a Canadian producer of devices that "convert computer generated print files directly to plates used for printing," represents a vertical acquisition. Because Kodak's sales of traditional film and developing have been declining as more people turn to digital photography, Kodak has been acquiring firms that move it into the "filmless imaging" area. These acquisitions have included digital printing (as in the Creo acquisition), health-care imaging, and consumer photography markets. Although the Creo acquisition is primarily focused on a corporate market, other acquisitions will allow Kodak to sell imaging products across a range of specialty (e. g. , health care) and consumer markets.

Vertical acquisitions also occur in service and entertainment businesses. Sony's acquisition of Columbia Pictures in the late 1980s was a vertical acquisition in which Columbia's movie content could be used by Sony's hardware devices. Sony's additional acquisition of CBS Records, a music producer, and development of the PlayStation hardware have formed the bases for more vertical integration. The spread of broadband and the technological shift from analog to digital hardware require media firms to find new ways to sell their content to consumers. Sony's former CEO, Nobuyuki Idei, believed that this shift created a new opportunity to sell hardware that integrates this change by selling "televisions, personal computers, game consoles and handheld devices through which all of that wonderful content will one day be streaming. "

However, this vision has not functioned well, and Idei was replaced by Howard Stringer as CEO, the first American CEO in Sony's history. Sony's businesses were quite autonomous and the coordination proved difficult to establish between them to realize Idei's vision. Furthermore, the lack of coordination caused a slowdown in innovation such that "Sony's reputation as an innovator" has suffered as "the snazziest gadgets from competitors, like the iPod and the TiVo digital video recorder, increasingly depend on the specific juggling act that Sony can't do well: integrating hardware, software and services. "

Barriers to entry are factors associated with the market or with the firms currently operating in it that increase the expense and difficulty faced by new ventures trying to enter that particular market. For example, well-established competitors may have substantial economies of scale in the manufacture or service of their products. In addition, enduring relationships with customers often create product loyalties that are difficult for new entrants to overcome. When facing differentiated products, new entrants typically must spend considerable resources to advertise their goods or services and may find it necessary to sell at prices below competitors' to entice customers. Facing the entry barriers created by economies of scale and differentiated products, a new entrant may find acquiring an established company to be more effective than entering the market as a competitor offering a good or service that is unfamiliar to current buyers. In fact, the higher the barriers to market entry, the greater the probability that a firm will acquire an existing firm to overcome them. Although an acquisition can be expensive, it does provide the new entrant with immediate market access.

For example, Nortel Networks Corp., a Canadian telecom producer, recently purchased PEC Solutions for $448 million. Through this acquisition, the new subsidiary, called Nortel PEC Solutions, inherited government contracts in the growing market pertaining to homeland security, intelligence, and defense. This gives Nortel a stronger stake in the federal computer networks market. Although other federal programs have been cut, the budget for information technology has increased from the proposed $60 billion in 2005 to $65 billion in 2006. Before the purchase, only 40 of Nortel's 30,000 employees worldwide had security clearances from the U. S. government. Nortel's purchase of PEC significantly increased this number, allowing Nortel to overcome considerable barriers to entry in this growing market. Furthermore, the combined company allows Nortel PEC to compete with the nation's largest contractors such as Lockheed Martin and Northrup Grumman. The acquisition has allowed Nortel to transition into this government service market much more rapidly than it would have been able to without buying a current player in the market. It also has given Nortel improved access to a market for its "large-scale telecommunications equipment."

As in the Nortel example, firms trying to enter international markets often face steep entry barriers. However, acquisitions are commonly used to overcome those barriers. At least for large multinational corporations, another indicator of the importance of entering and then competing successfully in international markets is the fact that five emerging markets (China, India, Brazil, Mexico, and Indonesia) are among the 12 largest economies in the world, with a combined purchasing power that is already one-half that of the Group of Seven industrial nations (United States, Japan, Britain, France, Germany, Canada, and Italy). Furthermore, the emerging markets are among the fastest growing economies in the world.

Acquisitions are another means a firm can use to gain access to new products and to current products that are new to the firm. Compared with internal product development processes, acquisitions provide more predictable returns as well as faster market entry. Returns are more predictable because the performance of the acquired firm's products can be assessed prior to completing the acquisition. 48 For these reasons, extensive bidding wars and acquisitions are more frequent in high-technology industries.

Acquisition activity is also extensive throughout the pharmaceutical industry, where firms frequently use acquisitions to enter markets quickly, to overcome the high costs of developing products internally, and to increase the predictability of returns on their investments. The cost of bringing a new drug to market in 2005 was "pushing $900 million and the average time to launch stretched to 12 years." Interestingly, there was one large deal between pharmaceutical firms in 2004, the merger between French firms Sanofi Synthelabo and Avenus that created Sanofi-Avenus. This $67 billion deal accounted for most of the $77.5 billion total value of deals between pharmaceutical firms. Although merger activity continued in 2005, most deals were smaller, as many companies targeted small acquisitions to supplement market power and reinvigorate or create innovative drug pipelines. Usually it is larger biotech or pharmaceutical firms acquiring smaller biotech firms that have drug opportunities close to market entry.

As indicated previously, compared with internal product development, acquisitions result in more rapid market entries. Acquisitions often represent the fastest means to enter international markets and help firms overcome the liabilities associated with such strategic moves. Acquisitions provide rapid access both to new markets and to new capabilities. Using new capabilities to pioneer new products and to enter markets quickly can create advantageous market positions. Pharmaceutical firms, for example, access new products through acquisitions of other drug manufacturers. They also acquire biotechnology firms both for new products and for new technological capabilities. Pharmaceutical firms often provide the manufacturing and marketing capabilities to take the new products developed by biotechnology firms to the market. In early 2005, Pfizer, for example, agreed to acquire Angiosyn, Inc., a smaller biotech company, which has developed a promising drug to avoid blindness. The deal, valued near $527 million, could extend Pfizer's lead in drugs for eye diseases. This deal m "spotlights the interest among the largest pharmaceutical makers to purchasing fledgling biotech concerns."

Because the outcomes of an acquisition can be estimated more easily and accurately than the outcomes of an internal product development process, managers may view acquisitions as lowering risk. The difference in risk between an internal product development process and an acquisition can be seen in the results of Pfizer's strategy and that of its competitors described above. As with other strategic actions discussed in this book, the firm must exercise caution when using a strategy of acquiring new products rather than developing them internally. While research suggests that acquisitions have become a common means of avoiding risky internal ventures (and therefore risky R&D investments), they may also become a substitute for innovation. Thus, acquisitions are not a risk-free alternative to entering new markets through internally developed products.

Acquisitions are also used to diversify firms. Based on experience and the insights resulting from it, firms typically find it easier to develop and introduce new products in markets currently served by the firm. In contrast, it is difficult for companies to develop products that differ from their current lines for markets in which they lack experience. Thus, it is uncommon for a firm to develop new products internally to diversify its product lines. Using acquisitions to diversify a firm is the quickest and, typically, the easiest way to change its portfolio of businesses. For example, since 2002 Advanced Medical Optics Inc. (AMO) has used an acquisition strategy to develop a set of products and services that focus on the "vision care lifecycle." AMO provides contact lenses to customers in their teens, laser surgery to patients in their 30s and 40s, and post-cataract-surgery implantable lenses to seniors.

In 2004 AMO acquired Pfizer's surgical ophthalmology business. In early 2005, the firm acquired Quest Vision Technologies, Inc., which focuses on developing lenses for presbyopia, or nearsightedness, a common condition among people in their 40s that causes them to wear bifocals or reading glasses. Finally, in 2005 they closed a deal to acquire Visx, Inc., the leader in laser surgery treatment machines based on sales volume. For the present, it appears that AMO's related diversification strategy is creating value as the market has valued its acquisitions positively.

8.8　Implementing Internal Innovations

An entrepreneurial mind-set is required to be innovative and to develop successful internal corporate ventures. When valuing environmental and market uncertainty, which are key parts of an entrepreneurial mind-set, individuals and firms demonstrate their willingness to take risks to commercialize innovations. While they must continuously attempt to identify opportunities, they must also select and pursue the best opportunities and do so with discipline. Thus, employing an entrepreneurial mind-set entails not only developing new products and markets but also placing an emphasis on execution. Those with an entrepreneurial mind-set "engage the energies of everyone in their domain," both inside and outside the organization.

Having processes and structures in place through which a firm can successfully implement the outcomes of internal corporate ventures and commercialize the innovations is critical. Indeed, the successful introduction of innovations into the marketplace reflects implementation effectiveness. In the context of internal corporate ventures, processes are the "patterns of interaction, coordination, communication, and decision making employees use" to convert the innovations resulting from either autonomous or induced strategic behaviors into successful market entries. Organizational structures are the sets of formal relationships supporting organizational processes.

Effective integration of the various functions involved in innovation processes— from engineering to manufacturing and, ultimately, market distribution—is required to implement the incremental and radical innovations resulting from internal corporate ventures. Increasingly, product development teams are being used to integrate the activities associated with different organizational functions. Such integration involves coordinating and applying the knowledge and skills of different functional areas in order to maximize innovation. Effective product development teams also create value when they "pull the plug" on a project. Although ending a project is difficult, sometimes because of emotional commitments to innovation-based projects, effective teams recognize when conditions change in ways that preclude the innovation's ability to create value as originally anticipated.

8.9   Cross-Functional Product Development Teams

Cross-functional teams facilitate efforts to integrate activities associated with different organizational functions, such as design, manufacturing, and marketing. In addition, new product development processes can be completed more quickly and the products more easily commercialized when cross-functional teams work effectively. Using cross-functional teams, product development stages are grouped into parallel or overlapping processes to allow the firm to tailor its product development efforts to its unique core competencies and to the needs of the market.

Horizontal organizational structures support the use of cross-functional teams in their efforts to integrate innovation-based activities across organizational functions. Therefore, instead of being built around vertical hierarchical functions or departments, the organization is built around core horizontal processes that are used to produce and manage innovations. Some of the core horizontal processes that are critical to innovation efforts are formal; they may be defined and documented as procedures and practices. More commonly, however, these processes are informal: "They are routines or ways of working that evolve over time." Often invisible, informal processes are critical to successful innovations and are supported properly through horizontal organizational structures more so than through vertical organizational structures. Two primary barriers that may prevent the successful use of cross-functional teams as a means of integrating organizational functions are independent frames of reference of team members and organizational politics.

Team members working within a distinct specialization (e. g., a particular organizational function) may have an independent frame of reference typically based on common backgrounds and experiences. They are likely to use the same decision criteria to evaluate issues such as product development efforts as they do within their functional units. Research suggests that functional departments vary along four dimensions: time orientation, interpersonal orientation, goal orientation, and formality of structure. Thus, individuals from different functional departments having different orientations on these dimensions can be expected to perceive product development activities in different ways. For example, a design engineer may consider the characteristics that make a product functional and workable to be the most important of the product's characteristics. Alternatively, a person from the marketing function may hold characteristics that satisfy customer needs most important. These different orientations can create barriers to effective communication across functions.

Organizational politics is the second potential barrier to effective integration in cross-functional teams. In some organizations, considerable political activity may center on allocating resources to different functions. Inter unit conflict may result from aggressive competition for resources among those representing different organizational functions. This dysfunctional conflict between functions creates a barrier to their integration. Methods must be found to achieve cross-functional integration without excessive political conflict and without changing the basic structural characteristics necessary for task specialization and efficiency. Cross-functional teams facilitate efforts to integrate activities associated with different organizational functions, such as design, manufacturing, and marketing. In addition, new product development processes can be completed more quickly and the products more easily commercialized when cross-functional teams work effectively. Using cross-functional teams, product development stages are grouped into parallel or overlapping processes to allow the firm to tailor its product development efforts to its unique core competencies and to the needs of the market.

Horizontal organizational structures support the use of cross-functional teams in their efforts to integrate innovation-based activities across organizational functions. Therefore, instead of being built around vertical hierarchical functions or departments, the organization is built around core horizontal processes that are used to produce and manage innovations. Some of the core horizontal processes that are critical to innovation efforts are formal; they may be defined and documented as procedures and practices. More commonly, however, these processes are informal: "They are routines or ways of working that evolve over time." Often invisible, informal processes are critical to successful innovations and are supported properly through horizontal organizational structures more so than through vertical organizational structures. Two primary barriers that may prevent the successful use of cross-functional teams as a means of integrating organizational functions are independent frames of reference of team members and organizational politics.

Team members working within a distinct specialization (e.g., a particular organizational function) may have an independent frame of reference typically based on common backgrounds and experiences. They are likely to use the same decision criteria to evaluate issues such as product development efforts as they do within their functional units. Research suggests that functional departments vary along four dimensions: time orientation, interpersonal orientation, goal orientation, and formality of structure. Thus, individuals from different functional departments having different orientations on these dimensions can be expected to perceive product development activities in different ways. For example, a design engineer may consider the characteristics that make a product functional and workable to be the most important of the product's characteristics. Alternatively, a person from the marketing function may hold characteristics that satisfy customer needs most important. These different orientations can create barriers to effective communication across functions.

Organizational politics is the second potential barrier to effective integration in cross-functional teams. In some organizations, considerable political activity may center on allocating resources to different functions. Inter unit conflict may result from aggressive competition for resources among those representing different organizational functions. This dysfunctional conflict between functions creates a barrier to their integration. Methods must be found to achieve cross-functional integration without excessive political conflict and without changing the basic structural characteristics necessary for task specialization and efficiency.

8.10 The decision-making process

As you can see from this example, managers at all levels, and in all areas, of organizations make decisions. That is, they make choices from two or more alternatives. For instance, top-level managers make decisions about their organization's goals, where to locate manufacturing facilities, what new markets to move into, and what products or services to offer. Middle and lower-level managers make decisions about setting weekly or monthly production schedules, handling problems that arise, allocating pay rises, and selecting or disciplining employees. But making decisions is not something that only managers do. All organizational members make decisions that affect their jobs and the organization they work for. How do they make those decisions?

Although decision making is typically described as 'choosing between alternatives', that view is too simplistic. Why? Because decision making is a far more comprehensive process than this simple act implies. 2 Even for something as straightforward as deciding where to go for lunch, you do more than just choose between a hamburger and a sandwich. Granted, you may not spend a lot of time contemplating the lunch decision, but you still engage in the steps in the decision-making process. What does the decision-making process involve?

The assumptions of rationality apply to any decision. Because we are concerned with managerial decision making, however, we need to add one further assumption. Rational managerial decision making assumes that decisions are made in the best economic interests of the organization. That is, the decision maker is assumed to be maximizing the organization's interests, not their own interests.

How realistic are these assumptions about rationality? Managerial decision making can follow rational assumptions if the following conditions are met: the manager is faced with a simple problem in which the goals are clear and the alternatives limited, in which the time pressures are minimal and the cost of seeking out and evaluating alternatives is low, for which the organizational culture supports innovation and risk taking, and in which the outcomes are relatively concrete and measurable. But most decisions that managers face in the real world do not meet all those tests. How, then, are most decisions in organizations usually made? The concept of bounded rationality can help answer that question.

Despite the limits to perfect rationality, managers are expected to follow a rational process when making decisions. Managers know that 'good' decision makers are supposed to do certain things and exhibit good decision-making behaviors as they identify problems, consider alternatives, gather information, and act decisively but prudently. By doing so, they are signaling to their superiors, peers and subordinates that they are competent and that their decisions are the result of intelligent and rational deliberation. However, certain aspects of the decision-making process are not realistic with respect to how managers make decisions. Instead, managers tend to operate under assumptions of bounded rationality; that is, they make decisions rationally, but are limited (or bounded) by their ability to process information. Because they cannot possibly analyse all information on all alternatives, managers satisfice, rather than maximise. That is, they accept solutions that are 'good enough'. They are being rational within the limits (bounds) of their ability to process information. Let us look at an example.

Most decisions that managers make do not fit the assumptions of perfect rationality; instead, they satisfied. However, keep in mind that their decision making also may be strongly influenced by the organization's culture, internal politics and power considerations, and by a phenomenon called escalation of commitment, which is an increased commitment to a previous decision despite evidence that it may have been wrong. For example, studies of the events leading up to the Challenger space shuttle disaster point to an escalation of commitment by decision makers to launch the shuttle on that fateful day even though the decision was questioned by certain individuals. Why would decision makers want to escalate commitment to a bad decision? Because they do not want to admit that their initial decision may have been flawed. Rather than search for new alternatives, they simply increase their commitment to the original solution.

What role does intuition play in managerial decision making? Managers regularly use their intuition and it may actually help to improve their decision making. What is intuitive decision making? It is making decisions on the basis of experience, feelings and accumulated judgment. Making a decision on intuition or 'gut-feeling' does not necessarily happen independently of rational analysis; rather, the two complement each other. A manager who has had experience with a particular, or even similar, type of problem or situation often can act quickly with what appears to be limited information. Such a manager does not rely on a systematic and thorough analysis of the problem or identification and evaluation of alternatives but instead uses his or her experience and judgment to make a decision.

How common is intuitive decision making? One survey of corporate executives found that almost half of them used intuition more than formal analysis to run their companies. Intuitive decision making can complement both rational and bloodedly rational decision making. A manager who has had experience with a similar type of problem or situation often can act quickly with what appears to be limited information because of that past experience. Furthermore, a recent study has found that individuals who experience intense feelings and emotions when making decisions actually achieve higher decision-making performance, especially when they understand their feelings as they are making decisions. The old belief that managers should ignore emotions when making decisions may not be the best advice.

8.11   Information is a Product

Many companies find that the information they have about their customers is valuable not only to themselves, but to others as well. A supermarket with a loyalty card program has something that the consumer packaged goods industry would love to have—knowledge about who is buying which products. A credit card company knows something that airlines would love to know—who is buying a lot of airplane tickets. Both the supermarket and the credit card company are in a position to be knowledge brokers or infomediaries. The supermarket can charge consumer packaged goods companies more to print coupons when the supermarkets can promise higher redemption rates by printing the right coupons for the right shoppers. The credit card company can charge the airlines to target a frequent flyer promotion to people who travel a lot, but fly on other airlines.

Google knows what people are looking for on the Web. It takes advantage of this knowledge by selling sponsored links. Insurance companies pay to make sure that someone searching on "car insurance" will be offered a link to their site. Financial services pay for sponsored links to appear when someone searches on the phrase "mortgage refinance". In fact, any company that collects valuable data is in a position to become an information broker. The Cedar Rapids Gazette takes advantage of its dominant position in a 22-county area of Eastern Iowa to offer direct marketing services to local businesses. The paper uses its own obituary pages and wedding announcements to keep its marketing database current.

There is always a lag between the time when new algorithms first appear in academic journals and excite discussion at conferences and the time when commercial software incorporating those algorithms becomes available. There is another lag between the initial availability of the first products and the time that they achieve wide acceptance. For data mining, the period of widespread availability and acceptance has arrived. Data mining is being used to promote customer retention in any industry where customers are free to change suppliers at little cost and competitors are eager to lure them away. Banks call it attrition. Wireless phone companies call it churn. By any name, it is a big problem. By gaining an understanding of who is likely to leave and why, a retention plan can be developed that addresses the right issues and targets the right customers.

Signet acquired behavioral data from many sources and used it to build predictive models. Using these models, it launched the highly successful balance transfer program that changed the way the credit card industry works. In 1994, Signet spun off the card operation as Capital One, which is now one of the top 10 credit card issuers. The same aggressive use of data mining technology that fueled such rapid growth is also responsible for keeping Capital One's loan loss rates among the lowest in the industry. Data mining is now at the heart of the marketing strategy of all the major credit card issuers.

Credit card divisions may have led the charge of banks into data mining, but other divisions are not far behind. At Wachovia, a large North Carolina-based bank, data mining techniques are used to predict which customers are likely to be moving soon. For most people, moving to a new home in another town means closing the old bank account and opening a new one, often with a different company. Wachovia set out to improve retention by identifying customers who are about to move and making it easy for them to transfer their business to another Wachovia branch in the new location. Not only has retention improved markedly, but also a profitable relocation business has developed.

In addition to setting up a bank account, Wachovia now arranges for gas, electricity, and other services at the new location. These applications should give you a feel for what is possible using data mining, but they do not come close to covering the full range of applications. The data mining techniques described in this book have been used to find quasars, design army uniforms, detect second-press olive oil masquerading as "extra virgin," teach machines to read aloud, and recognize handwritten letters. They will, no doubt, be used to do many of the things your business will require to grow and prosper for the rest of the century. In the next chapter, we turn to how businesses make effective use of data mining, using the virtuous cycle of data mining.

Chapter 9   Strategic Evaluation and Control

Evaluation and control is the process by which corporate activities and performance results are monitored so that actual performance can be compared with desired performance. 评价和控制是监控企业活动的绩效结果的过程，有助于将实际绩效与预期绩效进行比较。

This process can be viewed as a five-step feedback model, as depicted in: determine what to measure; establish standards of performance; measure actual performance; compare actual performance with the standard; take corrective action. 这个过程被看作五个反馈的步骤，即确定衡量内容、建立绩效标准、衡量实际绩效表现、根据标准比较实际绩效表现、采取纠正措施。

9.1   Product Diversification as an Example of an Agency Problem

A corporate-level strategy to diversify the firm's product lines can enhance a firm's strategic competitiveness and increase its returns, both of which serve the interests of all stakeholders and certainly shareholders and top-level managers. However, product diversification can create two benefits for top-level managers that shareholders do not enjoy, meaning that they may prefer product diversification more than shareholders do.

The fact that product diversification usually increases the size of a firm and that size is positively related to executive compensation is the first of the two benefits of additional diversification that may accrue to top-level managers. Diversification also increases the complexity of managing a firm and its network of businesses, possibly requiring additional managerial pay because of this complexity for top-level managers to increase their compensation.

9.2   How Corporate Strategies Form

In diversified companies corporate strategy tends to emerge incrementally As internal and external events unfold As managers Probe the future Experiment Gather more information Sense problems Build awareness of options Spot new opportunities. Develop ad hoc responses to unexpected crises. Acquire a feel for strategically relevant factors and their importance and interrelationships Develop consensus of how to proceed Our strategy will be.

Managing the Process of Crafting Corporate Strategy Not done all at once in comprehensive fashion Approached a step at a time, emerging gradually Begin with broad, intuitive concepts and then fine-tune and embellish them as More information is gathered Formal analysis confirms or modifies emerging judgments about situation Confidence and consensus build for the proposed strategic moves.

9.3 Building Core Competencies and Competitive Capabilities

Implementing and Executing Strategy Action-oriented, operations-driven activity revolving around managing people and business processes Tougher and more time-consuming than crafting strategy. Success depends on doing a good job of Leading Motivating Working with others Creating fits between requirements for good strategy execution and how organization conducts its business Implementation involves.

Why Implementing and Executing Strategy is a Tough Management Job Demanding variety of managerial activities that have to be performed Numerous ways to tackle each activity Requires good people management skills Requires launching and managing a variety of initiatives simultaneously Number of be deviling issues to be worked out Battling resistance to change. Hard to integrate efforts of many different work groups into a smoothly-functioning whole.

Implementing a Newly Chosen Strategy Requires Adept Leadership Implementing a new strategy takes adept leadership to Convincingly communicate reasons for the new strategy Overcome pockets of doubt Build consensus and enthusiasm Secure commitment of concerned parties Get all implementation pieces in place and coordinated.

Characteristics of the Strategy Implementation Process Every manager has an active role No 10-step checklists Few concrete guidelines Least charted, most open-ended part of strategic management Cuts across many aspects of "how to manage".

Characteristics of the Strategy Implementation Process (continued). Each implementation situation occurs in a different context, affected by differing Business practices and competitive situations Work environments and cultures Policies Compensation incentives Mix of personalities and firm histories. Approach to implementation has be customized to fit the situation People implement strategies-Not companies!

The Eight Components of Implementing and Executing Strategy Building a Capable Organization Allocating Resources Establishing Strategy-Supportive Policies Instituting Best Practices for Continuous Improvement Installing Support ying Rewards Systems to Achievement of Key Strategic Targets Exercising Strategic Leadership Shaping Corporate Culture to Fit Strategy Strategy Implementer's Action Agenda.

9. 4   Key Traits to Building Core Competencies

Rarely grounded in skills or know-how of a single department. Typically emerge from collaborative efforts of different work groups, requiring senior management oversight. Leveraging competencies into competitive advantage requires concentrating more effort and talent than rivals on strengthening competencies to create valuable organizational capabilities. Sustaining competitive advantage requires adaptingcompetencies to new conditions

Developing CompetitivelyValuable Competencies Involves Managing human skills, knowledge bases, and intellect Coordinating efforts of related work groups Collaborative networking among internal groups and with external partners. Achieving dominating depth Senior managers have to guide the process Ongoing challenge: Broaden, deepen, or modify competencies and capabilities in response to market changes.

Building Competencies: Keys to Success Selecting superior employees. Training Cultural influences Cooperation and collaboration Motivation Empowerment Attractive incentives Organizational flexibility Short deadlines Good databases.

The Most Valuable Organizational Capabilities Contribute heavily to better strategy execution. Provide a differentiating factor customers can see and value Difficult for rivals to match Time consuming to build Difficult to purchase Hard to replicate or imitate.

Process of BuildingOrganizational Capabilities. Develop ability to do something. Select people with relevant skills/experience Upgrade individual abilities as needed Mold work of employees into cooperative effort. As experience builds, ability can translate into a competence and/or capability. Capability becomes a distinctive competence, resulting in a potential competitive advantage

Process of BuildingOrganizational Capabilities: Step 1Develop ability to do something Select people with relevant skills/experience Broaden or deepen individual abilities as needed Mold efforts and work products of individuals into a cooperative group effort to create organizational ability.

Process of Building Organizational Capabilities: Step 2. As experience builds, such that the organization learns to accomplish the activity consistently well and at acceptable cost, the "ability" begins to translate into a competence and/or a capability Capabilities emerge from establishing and nurturing collaborative working relationships between individuals and groups in departments and between a company and its external allies.

Process of Building Organizational Capabilities: Step 3. If mastery is achieved to the point where the organization has the capability to perform the activity better than rivals, the "capability" becomes a distinctive competence and holds potential for competitive advantage The optimal outcome of the capability-building process!

Updating Competencies and Capabilities as Conditions Change Competencies and capabilities must continuously be modified and perhaps seven replaced with new ones due to. New strategic requirements Evolving market conditions Changing customer expectations Ongoing efforts to keep core competencies up-to-date can provide a basis for sustaining both Effective strategy execution and Competitive advantage.

Strategic Role of Employee Training Plays a critical role in implementation when a firm shifts to a strategy requiring different Skills-based competencies Competitive capabilities Managerial approaches Operating methods Types of training approaches Internal "universities" Orientation sessions for new employees Tuition reimbursement programs Online training courses.

Matching Organization Structure to Strategy Few hard and fast rules for organizing The One Big Rule: The role and purpose of the organization structure is to support and facilitate good strategy execution! Each firm's structure is idiosyncratic, reflecting Prior arrangements and internal politics. Executive judgments and preferences about how to arrange reporting relationships. How best to integrate and coordinate work effort of different work groups and departments Vice President Vice President Vice President CEO.

Structuring the Organization to Promote Successful Strategy Execution. Identify strategy-critical value chain activities Decide value chain activities to perform internally and those to outsource Make internally-performed strategy-critical value chain activities the main building blocks in the structure. Decide how much authority to centralize at the top and how much to delegate to managers and employees. Provide cross-unit coordination and collaboration to build/strengthen internal competencies and capabilities Provide the necessary collaboration and coordination with outsiders.

Step 1: Identify Strategy-Critical Activities Which activities are strategy-critical depends on Particulars of a firm's strategy Value-chain make-up Competitive requirements. External market conditions. Identify strategy-critical activities1. What business processes have to be performed extra well or in timely fashion to achieve competitive advantage? 2. In what value-chain activities would poor work performance impair strategic success Critical activities.

Step 2: Potential Advantages of Outsourcing Non-Critical Activities Decrease internal bureaucracies Flatten organization structure Speed decision-making. Provide firm with heightened strategic focus Improve a firm's innovative capacity. Increase competitive responsiveness Outsourcing makes strategic sense when outsiders can perform certain activities at a lower cost and/or with higher value-added.

Appeal of Outsourcing non-critical activities allows a firm to concentrate its energies and resources on those value-chain activities where it Can create unique value Can be best in the industry Needs strategic control to Build core competencies Achieve competitive advantage Manage key customer-supplier—distributor relationships.

Potential Advantages of Partnering By building, improving, and then leveraging partnerships, a firm enhances its overall capabilities and builds resource strengths that Deliver value to customers Rivals can't quite match. Consequently pave the way for competitive success Partnering makes strategic sense when the result is to enhance organizational capabilities.

Step 3: Make Strategy-Critical Activities the Main Building Blocks Assign managers of strategy-critical activities a visible, influential position Avoid fragmenting responsibility for strategy-critical activities across many departments. Provide coordinating linkages between related work groups Meld into a valuable competitive capability Assign managers key roles Primary activities Strategic relationships CoordinationValuable capability Support functions.

Strategic Management Principle Matching structure to strategy requires making strategy-critical activities and organizational units the main building blocks in the organization structure!

Why Structure Follows Strategy Changes in strategy typically require a new structure New strategy often involves different skills, different key activities, different staffing and organizational requirements Hence, a new strategy signals a need to reassess the organization structure How work is structured is a means to an end—not an end in itself!

Guard Against Functional Designs That Fragment Activities Scattering pieces of critical business processes across several specialized departments results in Many hand-offs which Lengthens completion time Increases coordination and overhead costs. Increases risk of details falling through the cracks Obsession with activity rather than result Solution Business process reengineering. Involves pulling strategy-critical processes from functional silos to create process-complete departments or cross- functional work groups.

Example: Fragmented Strategy-Critical Activities in a Functional Structure Filling customer orders Speeding new products to market Improving product quality Supply chain management Building capability to conduct business via the Internet Obtaining feedback from customers, making product modifications to meet their needs,

Step 4: Determine How Much Authority to Delegate to Whom In a centralized structure Top managers retain authority for most decisions In a decentralized structure Managers and employees are empowered to make decisions Trend in most companies Shift from authoritarian to decentralized structures stressing empowerment.

Advantages of Decentralized Decision-Making and Empowerment Fewer management layers Less bureaucracy Shorter response times More creativity and new ideas Better motivation of employees Greater employee involvement Increased organizational capability.

Principles Underlying the Global Trend Toward Decentralization and Empowerment. As the world economy moves into the Internet Age, traditional hierarchical structures must undergo radical surgery to capitalize on External market and Internal operating potential of e-commerce. Decisions are best made at the lowest organizational level capable to make timely, informed, competent decisions. Empowering employees to exercise judgment on job-related matters improves motivation and job performance.

Step 5: Reporting Relationships and Cross-Unit Coordination Classic method of coordinating activities-Have related units report to single manager Upper-level managers have clout to coordinate/unify efforts of their units. Support activities should be woven into structure in ways to Maximize performance of primary activities Contain costs of support activities Formal reporting relationships often need to be supplemented.

Options to Supplement the Basic Organization Structure Coordinating teams Cross-functional task forces Dual reporting relationships Informal networking Incentive compensation tied to group performance Teamwork and inter-departmental cooperation.

Step 6: Assign Responsibility for Collaboration With Outsiders Need multiple ties at multiple levels to ensure CommunicationCoordination and control Find ways to produce collaborative efforts to enhance firm's capabilities and resource strengths While collaborative relationships present opportunities, nothing valuable is realized until the relationship develops into an engine for better organizational performance.

Roles of Relationship Managers With Strategic Partners Get the right people together Promote good rapport See that plans for specific activities are developed and implemented. Help adjust internal procedures and communication systems Successfully link partners Iron out operating dissimilarities Nurture interpersonal ties.

Perspectives on Organizing All basic organization designs have strategy-related strengths and weaknesses. No ideal organization design exists To do a good job of matching structure to strategy Pick a basic design Modify as needed Supplement with appropriate coordinating, networking, and communication mechanisms to support effective execution of the strategy.

Organizational Structures of the Future: Overall Themes Revolutionary changes in how companies organize work have been triggered by New strategic priorities Rapidly shifting competitive conditions Tools of organizational design include Empowered managers and workers Reengineered work processes Self-directed work teams. Rapid incorporation of Internet technologies and cutting-edge e-commerce infrastructure Networking with outsiders.

Organizational Structures of the Future: Overall Themes Traditional, authoritarian structures have often proved to be a liability where Market conditions are fluid Customer preferences shift from standardized to customized products Product life-cycles grow shorter

Flexible manufacturing replaces mass production Customers want to be treated as individuals Pace of technological change accelerates.

Organizational Structures of the Future: Requirements for Success Decentralized structures with fewer managers Small-scale business units Reengineering to decrease fragmentation Development of stronger and newer capabilities Collaborative partnerships with outsiders Empowerment and self-directed work teams Lean staffing of corporate support functions Electronic information systems Accountability for results Use of e-commerce in daily operations.

Characteristics of Organizations of the Future The future structure will be. Change&Learning Fewer boundaries between Different vertical ranks Functions and disciplines Units in different geographic locations Firm and its suppliers, distributors, strategic allies, and customers Capacity for change and learning Collaborative efforts among people indifferent functions and geographic locations Extensive use of e-commerce technology and Internet business practices.

9.5   Instituting Best Practices and Installing Support Systems

Linking Budgets to Strategy Allocating resources in ways that support effective strategy execution involves Funding capital projects that can make a contribution to strategy implementation Funding efforts to strengthen competencies and capabilities or to create new ones Shifting resources—downsizing some areas, upsizing others, killing activities no longer justified, and funding new activities with a critical strategy role.

Strategic Management Principle Depriving strategy-critical groups of the funds needed to execute their pieces of the strategy can undermine the implementation process!

How Policies and Procedures Aid Strategy Implementation Provide top-down guidance regarding expected behaviors Help align internal actions with strategy, channeling efforts along the intended path. Enforce consistency in performance of activities in geographically scattered units Serve as powerful lever for changing corporate culture to produce stronger fit with a new strategy.

Creating Strategy-Supportive Policies and Procedures Role of new policies Channel behaviors and decisions to promote strategy execution Counteract tendencies of people to resist chosen strategy Too much policy can be as stifling as Wrong policy or as Chaotic as no policy. Often, the best policy is empowering employees and letting them operate between the white lines anyway they think best.

Instituting Best Practices and Continuous Improvement Searching out and adopting best practices is integral to effective implementation Benchmarking has spawned new approaches to improve strategy execution Reengineering TQM Continuous improvement programs.

Characteristics of Benchmarking Involves determining how well a firm performs particular activities and processes against "Best in industry" and/or "Best in world" performers Represents a solid methodology to identify options to improve Caution-Exact duplication of best practices of other firms is not feasible due to differences in implementation situations. Best approach-Best practices of other firms need to be modified or adapted to a firm's own specific situation Best Practices.

Information evaluation and control consist of performance data and activity reports. 信息评价和控制信息涉及性能数据和活动报告。

Information evaluation and control must be relevant to what is being monitored. 信息评价和控制必须与监测有关。

Evaluation and control is not an easy process. One of the obstacles to effective control is the difficulty in developing appropriate measures of important activities and outputs. 评价和控制并非那么容易的过程。原因之一是在重要的活动和表现中发展合适措施的有效控制是困难的。

Corporations are increasingly being evaluated on criteria other than economic. 目前越来越多的公司评估标准已经不仅仅是经济指标。

9.6 What Total Quality Management Is

TQM is a philosophy of managing a set of business practices that emphasizes Continuous improvement in all phases of operations，100 percent accuracy in performing activities，Involvement and empowerment of employees a tall levels，Team-based work design，Benchmarking，and Fully satisfying customer expectations. Goals of Quality Improvement Programs Defect-free manufacture Superior product quality Superior customer service Total customer satisfaction.

Components of Popular TQM Approaches Deming's 14 Points1. Constancy of purpose2. Adopt the philosophy3. Don't rely on mass inspection4. Don't award business on price5. Constant improvement6. Training7. Leadership8. Drive out fear9. Break down barriers10. Eliminate slogans and exhortations11. Eliminate quotas12. Pride of workmanship13. Education and retraining14. Plan of action.

Components of Popular TQM Approaches The Juran Trilogy Quality Planning Quality Control Quality Improvement? Set goals? Identify customers and their needs? Develop products and processes? Evaluate performance? Compare to goal sand adapt? Establish infrastructure? Identify projects and teams? Provide resources and training? Establish controls.

Components of Popular TQM Approaches Crosby's Quality Steps. Management commitment. Quality improvement teams. Quality measurement. Cost of quality evaluation. Quality awareness. Corrective action. Zero-defects committee. Supervisor training. Zero-defects day. Goal-setting. Error cause removal. Recognition. Quality councils. Do it over again.

Components of Popular TQM Approaches 1992 Baldridge Award Criteria(1000 points)Quality1. Leadership(90 points)2. Information &analysis(80 points)3. Strategic quality planning(60 points) 4. Human resource development (150 points) 5. Management of process quality (140 points) 6. Quality&operation results (180 points) 7. Customer focus&satisfaction(300points).

Twelve Aspects Common to TQM and Continuous Improvement Programs1. Committed leadership2. Adoption and communication of TQM3. Closer customer relationships4. Closer supplierrelationships5. Benchmarking6. Increased training7. Open organization8. Employee empowerment9. Zero-defects mentality10. Flexible manufacturing11. Process improvement12. Measurement.

Implementing a Philosophy of Continuous Improvement In still enthusiasm to do things right throughout company Strive to achieve little steps forward each day, (what the Japanese call kaizen). Ignite creativity in employees to improve performance of value-chain activities. Preach there is no such thing as good enough Reform the corporate culture.

Characteristics of TQM/ContinuousImprovement Programs Valuable competitive asset in a company's resource portfolio Have hard-to-imitate aspects Require substantial investment of management time and effort Expensive in terms of training and meetings Seldom produce short-term results Long-term payoff—instilling a TQM culture.

TQM vs. Process Reengineering Aims at quantum gains of 30 to 50%or more TQM Stresses incremental progress. Techniques are not mutually exclusive Reengineering-Used to produce a good basic design yielding dramatic improvements TQM-Used to perfect process, gradually improving efficiency and effectiveness.

Using Best Practice Programs as an Implementation Tool Select indicators of successful strategy execution Benchmark against best practice companies Reengineer business processes. Build a TQ culture Requires top management commitment Install TQ-supportive employee practices. Empower employees to do the right things Provide employees with quick access to required information Preach that performance can be improved.

Installing Support Systems Essential to promote successful strategy execution Types of support systems On-line data systems Internet and company intranets. Electronic mail E-commerce systems Mobilizing information and creating systems touse knowledge effectively can yield Competitive advantage.

Examples: Support Systems Airlines Computerized reservation system Federal

Express Computerized parcel-tracking system, leading-edge flight operations systems, and e-business tools.

Examples: Support Systems Otis Elevator Sophisticated maintenance support system Arthur Andersen Internet and digital technology (Knowledge Xchange system has data, voice, and video capabilities) links more than 70, 000 people in 382 offices in 81 countries.

Examples: Support Systems Domino's Pizza Computerized systems at each outlet facilitate ordering, inventory, payroll, cash flow, and work flow functions Mrs. Fields' Cookies System to monitor sales, at 15-minute intervals, to suggest product mix changes and to improve customer response.

Strategic Management Principle Innovative, state-of-the-art support systems can be a basis for competitive advantage if they give a firm capabilities that rivals can't match!

Formal Reporting of Strategy-Critical Information Accurate, timely information is essential to guide action Prompt feedback on implementation activities is needed before actions are fully completed Key strategic performance indicators must be tracked as often as practical Barometers of overall performance Statistical information Reports and meetings Personal contact.

9.7　What Areas Information Systems Should Address

Customer data Operations data Employee data Supplier/partner/collaborative ally data Financial performance data.

Exercising Adequate Control Over Empowered Employees Challenge How to ensure actions of employees stay within acceptable bounds Purpose of diagnostic control systems Relieve managers of burden of constant monitoring Control methods. Establish boundaries on what not to do, allowing freedom to act with limits Face-to-face meetings to assess performance.

Gaining Commitment: Components of an Effective Reward System Monetary Incentives Salary raises Performance bonuses Stock options Retirement packages Promotions Perks Non-monetary Incentives Praise Constructive criticism Special recognition More, or less, job security Interesting assignments More, or less, job responsibility.

Approaches: Motivating People to Execute the Strategy Well Inspire employees to do their best Get employees to buy into strategy Structure individual efforts in teams to facilitate a supportive climate. Allow employees to participate in decisions about their jobs Make jobs interesting and satisfying Devise strategy-supportive motivational approaches.

Examples: Motivational Practices No Lay-Off Policies Japanese automobile producers, along with several U. S. based companies (Southwest Airlines, FedEx, Lands'End, and

Harley Davidson) have no lay-off policies, using employment security both as a positive motivator and a means of reinforcing good strategy execution.

Examples: Motivational Practices Stock Options More than 35 of the 58 publicly held companies on the 1999 list of the 100 Best Companies to Work for in America (includes Cisco Systems, Procter&Gamble, Merck, Charles Schwab, General Mills, Amgen, and Tellabs) provide stock options to all employees. Having employee-owners sharing in a company's success is widely viewed as a positive motivator.

Examples: Motivational Practices Nordstrom Pay sales people higher than prevailing rates, plus commission. "Rule1: Use good judgment in all situations. There will be no additional rules. " Cisco Systems Offers on-the-spot bonuses of up to $2, 000 for exceptional performance.

Examples: Motivational Practices Microsoft Team members enjoy working 60-80hours per week for a leading edge company, accompanied by attractive pay and lucrative stock options. Lincoln Electric Rewards productivity by paying for each piece produced (defects can be traced to worker causing them). Bonuses of 50 to 100% are common.

Balancing Positive vs. Negative Rewards Elements of both are necessary Challenge and competition are necessary for self-satisfaction Prevailing view Positive approaches work better than negative ones in terms of Enthusiasm Effort Creativity Initiative.

Linking the Reward System to Performance Outcomes Rewards are the single most powerful tool to win commitment to the strategy Objectives Generously reward those achieving objectives Deny rewards to those who don't Make strategic performance measures the dominant basis for designing incentives, evaluating efforts, and handing out rewards.

Strategic Management Principle A properly designed reward structure is management's most powerful tool for mobilizing organizational commitment to successful strategy execution!

9.8 Guidelines for Designing an Effective Compensation System

Strategic Management Principle The unwavering standard for judging whether individuals, teams, and organizational units have done a good job must be whether hey achieve performance targets consistent with effective strategy execution!

Key Considerations in Designing Reward Systems Create a results-oriented system Reward people for results, not for activity Define jobs in terms of what to achieve Incorporate several performance measures Tie incentive compensation to relevant outcomes. Top executives—Key measures of overall firm performance Department heads, teams, and individuals Incentives tied to achieving performance targets in their areas of responsibility.

Payoff must be a major, not minor, piece of total compensation package. Incentive plan should extend to all employees. Administer system with scrupulous fairness. Link incentives to achieving only the performance targets in strategic plan. Targets each person is expected to achieve must involve outcomes that can be personally affected. Keep time between performance review and payment short Make liberal use of non-monetary rewards. Avoid ways of rewarding non-performers.

9.9　Controlling Process

It is the process of monitoring, comparing and correcting work performance. All managers should be involved in the control function even if their units are performing as planned. Managers cannot really know whether their units are performing properly until they have evaluated what activities have been done and have compared the actual performance with the desired standard. An effective control system ensures that activities are completed in ways that lead to the attainment of the organization's goals. The criterion that determines the effectiveness of a control system, then, is how well it helps employees and managers achieve their goals.

Why is control so important? Planning can be done, an organizational structure can be created to facilitate the efficient achievement of goals, and employees can be motivated through effective leadership. Still, there is no assurance that activities are going as planned and that the goals managers are seeking are, in fact, being attained. Control is important, therefore, because it is the final link in the management functions. It is the only way managers know whether organizational goals are being met and, if not, the reasons why. The value of the control function can be seen in three specific areas: planning, empowering employees and protecting the workplace.

The control process is a three-step process that involves measuring actual performance, comparing actual performance against a standard, and taking managerial action to correct deviations or inadequate standards. The control process assumes that performance standards already exist. These standards against which performance progress is measured are the specific goals created during the planning process. What we measure is probably more critical to the control process than how we measure. Why? Because selecting the wrong criteria can create serious problems. Besides, what is measured often determines what employees will do. What control criteria might managers use?

Some control criteria can be used for any management situation. For instance, because all managers, by definition, coordinate the work of others, criteria such as employee satisfaction or turnover and absenteeism rates can be measured. Most managers also have budgets set in dollar costs for their area of responsibility. Keeping costs within budget is, therefore, a fairly common control measure. However, any comprehensive control system needs to recognize the diversity of activities that managers perform. For instance, a manager at a pizza delivery location might use measures such as number of pizzas delivered per day, average delivery time, or number of coupons redeemed. On the other hand, the manager of an administrative unit in a governmental agency might use number of document pages typed per day, number of client requests completed per hour, or average time required to process paperwork. Marketing managers often use measures such as percentage of market held, average dollar per sale, number of customer visits per salesperson, or number of customer impressions per advertising medium.

Most work activities can be expressed in quantifiable and measurable terms. However, when a performance indicator cannot be stated in quantifiable terms, managers should use subjective measures. Although subjective measures have significant limitations, they are better than having no standards at all and doing no controlling. If an activity is important, the excuse that it is difficult to measure is unacceptable.

The comparing step determines the degree of variation between actual performance and the standard. Although some variation in performance can be expected in all activities, it is critical to determine the acceptable range of variation. Deviations that exceed this range become significant and need the manager's attention. In the comparison stage, managers are particularly concerned with the size and direction of the variation. An example should make this concept clearer.

Sports coaches understand the importance of correcting actual performance. During a game, they will often correct a player's actions. But if the problem is recurring or encompasses more than one player, they will devote time during practice before the next game to correcting the actions. That is what managers need to do as well. Depending on what the problem is, a manager could take different corrective actions. For instance, if unsatisfactory work is the reason for performance variations, the manager could correct it by things such as training programs, disciplinary action, changes in compensation practices, and so forth.

A manager who decides to correct actual performance has to make another decision: should immediate or basic corrective action be taken? Immediate corrective action corrects problems at once to get performance back on track. Basic corrective action looks at how and why performance has deviated and then proceeds to correct the source of the deviation. It is not unusual for managers to rationalize that they do not have the time to find the source of a problem (basic corrective action) and continue to perpetually 'put out fires' with immediate corrective action. Effective managers, however, analyze deviations and, when the benefits justify it, take the time to pinpoint and correct the causes of variance.

To return to the Cowley Wines New Zealand example, Mike Cowley might take immediate corrective action on the negative variance for Wynns Coonawarra Estate Shiraz. He might contact the wine producer and try to negotiate a 5 per cent reduction in price. However, taking basic corrective action would involve a more in-depth analysis. After assessing how and why sales deviated, Mike might increase in-store promotional efforts, increase the advertising budget for this brand, or reduce future orders with the wine producer. The action Mike takes will depend on the assessment of each brand's potential profitability.

It is possible that the variance was a result of an unrealistic standard; that is, the goal may have been too high or too low. In such cases, it is the standard that needs corrective attention, not the performance. In the example, Mike might need to raise the sales goal (standard) for Rosemount Diamond Label Shiraz to reflect its growing popularity. If performance consistently exceeds the goal, then a manager should look at whether the goal is too easy and needs to be raised.

On the other hand, managers must be cautious about revising a standard downward. It is natural to blame the goal when an employee or a team falls short. For instance, students who get a low grade on a test often attack the grade cut-off standards as being too high. Rather than accept the fact that their performance was inadequate, students argue that the standards are unreasonable.

Similarly, salespeople who fail to meet their monthly quota may attribute the failure to an unrealistic quota. The standard may well be too high, and if this is the case it can result in a significant variation and may even contribute to demotivating those employees being measured against it. But keep in mind that if employees or managers do not meet the standard, the first thing they are likely to attack is the standard. If you believe that the standard is realistic, fair and achievable, hold your ground. Explain your position; reaffirm to the employee, team or unit that you expect future performance to improve and then take the necessary corrective action to help make that happen.

9.10   Focus on organizational behavior

Organizational behavior focuses on two main areas. First, it looks at individual behavior. Based predominantly on contributions from psychologists, this area includes such topics as attitudes, personality, perception, learning and motivation. Second, OB is concerned with group behavior, which includes norms, roles, team building, leadership and conflict. Our knowledge about groups comes basically from the work of sociologists and social psychologists. Unfortunately, the behavior of a group of employees cannot be understood by merely summing up the actions of the individuals in the group, because individuals in a group setting behave differently from individuals acting alone. Think about yourself and your group of closest friends. Have you done things with your group of friends that you would never do on your own? Most of us have. Because employees in an organization are both individuals and members of groups, we need to study them at two levels. This chapter provides the foundation for understanding individual behavior. Then, in the next chapter, we will introduce the basic concepts related to understanding group behavior.

As a result of the Hawthorne Studies, managers generalized that if their employees were satisfied with their jobs, then that satisfaction would translate to working hard. So, for a good part of the 20th century managers believed that happy workers were productive workers. Because it has not been that easy to determine whether job satisfaction caused job productivity or vice versa, some management researchers felt that belief was generally wrong. However, we can say with some certainty that the correlation between satisfaction and productivity is fairly strong. And when satisfaction and productivity information is gathered for the organization as a whole, we find that organizations with more satisfied employees tend to be more effective than organizations with fewer satisfied employees.

It seems logical to assume that job satisfaction should be a major determinant of an employee's OCB. Satisfied employees would seem more likely to talk positively about the organization, help others, and go above and beyond normal job expectations. Research suggests that there is a modest overall relationship between job satisfaction and OCB. But that relationship is tempered by perceptions of fairness. Basically, if you do not feel as though your supervisor, organizational procedures or pay policies are fair, your job satisfaction is likely to suffer significantly. However, when you perceive that these things are fair, you have more trust in your employer and are more willing to engage voluntarily in behaviors that go beyond your formal job requirements. Another factor that influences individual OCB is the type of citizenship behavior a person's work group exhibits. In a work group with low group-level OCB, any individual in that group who engages in OCB is likely to have higher job performance ratings. One possible explanation may be that the person is trying to find some way to 'stand out' from the crowd. No matter why it happens, the point is that OCB can have positive benefits for organizations.

Job involvement is the degree to which an employee identifies with his or her job, actively participates in it, and considers his or her job performance to be important to his or her self-worth. Employees with a high level of job involvement strongly identify with and really care about the kind of work they do, as can be seen in the chapter-opening story about firefighters. Their generally positive attitude leads them to contribute in positive ways to their work. What do we know about the influence of job involvement on employee behavior? High levels have been found to be related to fewer absences, lower resignation rates and higher employee engagement with their work.

Organizational commitment is the degree to which an employee identifies with a particular organization and its goals, and wishes to maintain membership in the organization. Whereas job involvement is identifying with your job, organizational commitment is identifying with your employing organization. Research suggests that organizational commitment also leads to lower levels of both absenteeism and turnover and, in fact, is a better indicator of turnover than job satisfaction. Why? Probably because it is a more global and enduring response to the organization than is satisfaction with a particular job. However, organizational commitment is less important as a work-related attitude than it once was. Employees do not generally stay with a single organization for most of their career and the relationship they have with their employer has changed considerably. Although the commitment of an employee to an organization may not be as important as it once was, research about perceived organizational support - employees' general belief that their organization values their contribution and cares about their well-being-shows that the commitment of the organization to the employee can be beneficial. High levels of perceived organizational support lead to increased job satisfaction and lower turnover.

Chapter 10   Enterprise Growth Strategy

10. 1   Entrepreneurial Management

Building a startup is an exercise in institution building; thus, it necessarily involves management. This often comes as a surprise to aspiring entrepreneurs, because their associations with these two words are so diametrically opposed. Entrepreneurs are rightly of implementing traditional management practices early on in a startup, afraid that they will invite bureaucracy or stifle creativity. Entrepreneurs have been trying to the square peg of their unique problems into the round hole of general management for decades. As a result, many entrepreneurs take a "just do it" attitude, avoiding all forms of management, process, and discipline. Unfortunately, this approach leads to chaos more often than it does to success. I should know: my startup failures were all of this kind. The tremendous success of general management over the last century has provided unprecedented material abundance, but those management principles are ill suited to handle the chaos and uncertainty that startups must face.

We are living through an unprecedented worldwide entrepreneurial renaissance, but this opportunity is laced with peril. Because we lack a coherent management paradigm for new innovative ventures, we are throwing our excess capacity around with wild abandon. Despite this lack of rigor, we are some ways to make money, but for every success there are far too many failures: products pulled from shelves mere weeks after being launched, startups lauded in the press and forgotten a few months later, and new products that wind up being used by nobody. What makes these failures particularly painful is not just the economic damage done to individual employees, companies, and investors; they are also a colossal waste of our civilization's most precious resource: the time, passion, and skill of its people. The Lean Startup movement is dedicated to preventing these failures.

The Lean Startup takes its name from the lean manufacturing revolution that Taiichi Ohno and Shigeo Shingo are credited with developing at Toyota. Lean thinking is radically altering the way supply chains and production systems are run. Among its tenets are drawing on the knowledge and creativity of individual workers, the shrinking of batch sizes, just-in-time production and inventory control, and an acceleration of cycle times. It taught the world the between value-creating activities and waste and showed how to build quality into products from the inside out.

A comprehensive theory of entrepreneurship should address all the functions of an early-stage venture: vision and concept, product development, marketing and sales, scaling up, partnerships and distribution, and structure and organizational design. It has to provide a method for measuring progress in the context of extreme uncertainty. It can give entrepreneurs clear guidance on how to make the many decisions they face: whether and when to invest in process; formulating, planning, and creating infrastructure; when to go it alone and when to partner; when to respond to feedback and when to stick with vision; and how and when to invest in scaling the business. Most of all, it must allow entrepreneurs to make testable predictions.

Henry Ford is one of the most successful and celebrated entrepreneurs of all time. Since the idea of management has been bound up with the history of the automobile since its days, I believe it is setting to use the automobile as a metaphor for a startup. An internal combustion automobile is powered by two important and very t feedback loops. The feedback loop is deep inside the engine. Before Henry Ford was a famous CEO, he was an engineer. He spent his days and nights tinkering in his garage with the precise mechanics of getting the engine cylinders to move. Each tiny explosion within the cylinder provides the motive force to turn the wheels but also drives the ignition of the next explosion. Unless the timing of this feedback loop is managed precisely, the engine will sputter and break down.

Startups have a similar engine that I call the engine of growth. The markets and customers for startups are diverse: a toy company, a consulting, and a manufacturing plant may not seem like they have much in common, but, as we'll see, they operate with the same engine of growth. Every new version of a product, every new feature, and every Henry Ford is one of the most successful and celebrated entrepreneurs of all time. Since the idea of management has been bound up with the history of the automobile since its t days, I believe it is to use the automobile as a metaphor for a startup.

The second important feedback loop in an automobile is between the driver and the steering wheel. This feedback is so immediate and automatic that we often don't think about it, but it is steering that driving from most other forms of transportation. If you have a daily commute, you probably know the route so well that your hands seem to steer you there on their own accord. We can practically drive the route in our sleep. Yet if I asked you to close your eyes and write down exactly how to get to once—not the street directions but every action you need to take, every push of hand on wheel and foot on pedals—you'd it impossible. The choreography of driving is incredibly complex when one slows down to think about it.

By contrast, a rocket ship requires just this kind of in-advance calibration. It must be launched with the most precise instructions on what to do: every thrust, every ring of a booster, and every change in direction. The tiniest error at the point of launch could yield catastrophic results thousands of miles later. Unfortunately, too many startup business plans look more like they are planning to launch a rocket ship than drive a car. They prescribe the steps to take and the results to expect in excruciating detail, and as in planning to launch a rocket, they are set up in such a way that even tiny errors in assumptions can lead to catastrophic outcomes.

One company I worked with had the misfortune of forecasting customer adoption—in the millions—for one of its new products. Powered by a splashy launch, the company successfully executed its plan. Unfortunately, customers did not to the product in great numbers. Even worse, the company had invested in massive infrastructure, hiring, and support to handle the of customers it expected. When the customers failed to materialize, the company had committed itself so completely that they could not adapt in time. They had "achieved failure"—successfully, faithfully, and rigorously executing a plan that turned out to have been utterly flawed.

The Lean Startup method, in contrast, is designed to teach you how to drive a startup. Instead of making complex plans that are based on a lot of assumptions, you can make constant adjustments with a steering wheel called the Build-Measure-Learn feedback loop. Through this process of steering, we can learn when and if it's time to make a sharp turn called a pivot or whether we should persevere along our current path. Once we have an engine that's revved up, the Lean Startup methods to scale and grow the business with maximum acceleration.

Throughout the process of driving, you always have a clear idea of where you're going. If you're commuting to work, you don't give up because there's a detour in the road or you made a wrong turn. You remain thoroughly focused on getting to your destination. Startups also have a true north, a destination in mind: creating a thriving and world-changing business. I call that a startup's vision. To achieve that vision, startups employ a strategy, which includes a business model, a product road map, a point of view about partners and competitors, and ideas about who the customer will be. The product is the end result of this strategy (see the chart on this page). company had committed itself so completely that they could not adapt in time. They had "achieved failure"—successfully, faithfully, and rigorously executing a plan that turned out to have been utterly flawed.

10. 2　Related Diversification and Unrelated Diversification Strategies

Both related diversification and unrelated diversification strategies can be implemented through acquisitions. For example, United Technologies Corp. (UTC) has used acquisitions to build a conglomerate. Since the mid-1970s it has been building a portfolio of stable and non-cyclical businesses, including Otis Elevator Co. and Carrier Corporation (air conditioners), in order to reduce its dependence on the volatile aerospace industry. Its main businesses have been Pratt & Whitney (jet engines), Sikorsky (helicopters), and Hamilton Sundstrand (aerospace parts). UTC has also acquired a hydrogen-fuel-cell business. Perceiving an opportunity in security caused by problems at airports and because security has become a top concern both for governments and for corporations, United Technologies in 2003 acquired Chubb PLC, a British electronic-security company, for $1 billion. With its acquisition of Kidde PLC, in the same general business, in 2004 for $2.84 billion, UTC will have obtained 10 percent of the world's market share in electronic security. All businesses UTC purchases are involved in manufacturing industrial and commercial products.

However, many are relatively low technology (e. g. , elevators and air conditioners). Research has shown that the more related the acquired firm is to the acquiring firm, the greater the probability is that the acquisition will be successful. Thus, horizontal acquisitions (through which a firm acquires a competitor) and related acquisitions tend to contribute more to the firm's strategic competitiveness than would the acquisition of a company that operates in product markets quite different from those in which the acquiring firm competes.

10.3 Strategic entrepreneurship

Strategic entrepreneurship is taking entrepreneurial actions using a strategic perspective. When engaging in strategic entrepreneurship, the firm simultaneously focuses on finding opportunities in its external environment that it can try to exploit through innovations. Identifying opportunities to exploit through innovations is the entrepreneurship part of strategic entrepreneurship, while determining the best way to manage the firm's innovation efforts is the strategic part. Thus, strategic entrepreneurship finds firms integrating their actions to find opportunities and to successfully innovate as a primary means of pursuing them. In the 21st-century competitive landscape, firm survival and success increasingly is a function of a firm's ability to continuously find new opportunities and quickly produce innovations to pursue them.

To examine strategic entrepreneurship, we consider several topics in this chapter. First, we examine entrepreneurship and innovation in a strategic context. Definitions of entrepreneurship, entrepreneurial opportunities, and entrepreneurs as those who engage in entrepreneurship to pursue entrepreneurial opportunities are included as parts of this analysis. We then describe international entrepreneurship, a phenomenon reflecting the increased use of entrepreneurship in economies throughout the world. After this discussion, the chapter shifts to descriptions of the three ways firms innovate. Internally, firms innovate through either autonomous or induced strategic behavior. We then describe actions firms take to implement the innovations resulting from those two types of strategic behavior.

In addition to innovating through internal activities, firms can develop innovations by using cooperative strategies, such as strategic alliances, and by acquiring other companies to gain access to their innovations and innovative capabilities. Most large, complex firms use all three methods to innovate. The method the firm chooses to innovate can be affected by the firm's governance mechanisms. Research evidence suggests, for example, that inside board directors with equity positions favor internal innovation while outside directors with equity positions prefer acquiring innovation.

As you will see from studying this chapter, innovation and entrepreneurship are vital for young and old and for large and small firms, for service companies as well as manufacturing firms and for high-technology ventures. In the global competitive landscape, the long-term success of new ventures and established firms is a function of the ability to meld entrepreneurship with strategic management. Before moving to the next section, we should mention that our focus in this content is on innovation and entrepreneurship within established organizations. This phenomenon is called corporate entrepreneurship, which is the use or application of entrepreneurship within an established firm. An important part of the entrepreneurship discipline, corporate entrepreneurship increasingly is thought to be linked to survival and success of established corporations. Indeed, established firms use entrepreneurship to strengthen their performance and to enhance growth opportunities. Of course, innovation and entrepreneurship play a critical role in the degree of success achieved by start-up entrepreneurial ventures as well. Our focus in this chapter is on corporate entrepreneurship. However, the materials we will describe are equally important in entrepreneurial ventures (sometimes called "start-ups"). Moreover, we will make specific reference to entrepreneurial ventures in a few parts of the chapter as we discuss the importance of strategic entrepreneurship for firms competing in the 21st-century competitive landscape.

10. 4　Entrepreneurship and Entrepreneurial Opportunities

Entrepreneurship is the process by which individuals or groups identify and pursue entrepreneurial opportunities without being immediately constrained by the resources they currently control. Entrepreneurial opportunities are conditions in which new goods or services can satisfy a need in the market. These opportunities exist because of competitive imperfections in markets and among the factors of production used to produce them and when information about these imperfections is distributed asymmetrically (that is, not equally) among individuals. Entrepreneurial opportunities come in a host of forms (e. g. , the chance to develop and sell a new product and the chance to sell an existing product in a new market). Firms should be receptive to pursuing entrepreneurial opportunities whenever and wherever they may surface.

As these two definitions suggest, the essence of entrepreneurship is to identify and exploit entrepreneurial opportunities—that is, opportunities others do not see or for which they do not recognize the commercial potential. As a process, entrepreneurship results in the "creative destruction" of existing products (goods or services) or methods of producing them and replaces them with new products and production methods. Thus, firms engaging in entrepreneurship place high value on individual innovations as well as the ability to continuously innovate across time. We study entrepreneurship at the level of the individual firm. However, evidence suggests that entrepreneurship is the economic engine driving many nations' economies in the global competitive landscape. Thus, entrepreneurship, and the innovation it spawns, is important for companies competing in the global economy and for countries seeking to stimulate economic climates with the potential to enhance the living standard of their citizens.

Entrepreneurs are individuals, acting independently or as part of an organization, who see an entrepreneurial opportunity and then take risks to develop an innovation to pursue it. Often, entrepreneurs are the individuals who receive credit for making things happen! Entrepreneurs are found throughout an organization—from top-level managers to those working to produce a firm's goods or services. Entrepreneurs are found throughout W. L. Gore & Associates, for example. Recall from the Opening Case's analysis of this firm that part of the job of all Gore associates is to use roughly 10 percent of their time to develop innovations. Entrepreneurs tend to demonstrate several characteristics, including those of being optimistic, highly motivated, willing to take responsibility for their projects, and courageous. In addition, entrepreneurs tend to be passionate and emotional about the value and importance of their innovation-based ideas.

Evidence suggests that successful entrepreneurs have an entrepreneurial mind-set. The person with an entrepreneurial mind-set values uncertainty in the marketplace and seeks to continuously identify opportunities with the potential to lead to important innovations. Because it has the potential to lead to continuous innovations, individuals' entrepreneurial mind-sets can be a source of competitive advantage for a firm. Howard Schultz, founder of Starbucks, believes that his firm has a number of individuals with an entrepreneurial mind-set. Making music a meaningful part of Starbucks' customers' experiences is an example of an evolving product offering resulting from an entrepreneurial mind-set. In Schultz's words: "The music world is changing, and Starbucks and Starbucks

Hear Music will continue to be an innovator in the industry. It takes passion, commitment, and even a bit of experimentation to maintain that position. " Of course, changes in the music industry create the uncertainties that lead to entrepreneurial opportunities that can be pursued by relying on an entrepreneurial mind-set.

As our discussions have suggested, "innovation is an application of knowledge to produce new knowledge. " As such, entrepreneurial mind-sets are fostered and supported when knowledge is readily available throughout a firm. Indeed, research has shown that units within firms are more innovative when they have access to new knowledge. Transferring knowledge, however, can be difficult, often because the receiving party must have adequate absorptive capacity (or the ability) to learn the knowledge. This requires that the new knowledge be linked to the existing knowledge. Thus, managers need to develop the capabilities of their human capital to build on their current knowledge base while incrementally expanding that knowledge to facilitate the development of entrepreneurial mind-sets.

Recent actions at Hewlett-Packard Co. (HP) demonstrate the use of knowledge as part of the entrepreneurial mind-set many employees have developed. In response to inroads into the firm's lucrative computer printers being made by competitors such as Dell Inc. and Lexmark International, HP introduced a new technology for inkjet printers that reduces photo-printing time by half. An analyst commented, "This seems to be a pattern we've seen before—competitors gain on HP by slashing prices, then HP introduces new technology that lets them move ahead. "48 Thus, HP employees use their entrepreneurial mind-set to identify opportunities (e. g. , to reduce the time needed to print photos) and then integrate knowledge available throughout the firm to develop an innovation to exploit the identified opportunity.

10. 5   Incremental and Radical Innovation

Firms produce two types of internal innovations—incremental and radical innovations—when using their R&D activities. Most innovations are incremental—that is, they build on existing knowledge bases and provide small improvements in the current product lines. Incremental innovations are evolutionary and linear in nature. "The markets for incremental innovations are well-defined, product characteristics are well understood, profit margins tend to be lower, production technologies are efficient, and competition is primarily on the basis of price. " Adding a different kind of whitening agent to a soap detergent is an example of an incremental innovation, as are improvements in televisions over the last few decades (moving from black-and-white to color, improving existing audio capabilities, etc.). Panera Bread Company's introduction of new soups is another example of incremental innovations. Companies launch far more incremental innovations than radical innovations.

In contrast to incremental innovations, radical innovations usually provide significant technological breakthroughs and create new knowledge. Recall from the Opening Case that W. L. Gore & Associates seeks to develop primarily radical rather than incremental innovations through its R&D activities. Radical innovations, which are revolutionary and non-linear in nature, typically use new technologies to serve newly created markets. The development of the personal computer (PC) is an example of a radical innovation. Reinventing the computer by developing a "radically new computer-brain chip" is an example of what could be a radical innovation. If researchers are successful in their efforts, super chips (with the capability to process a trillion calculations per second) will be developed. Obviously, such a radical innovation would seem to have the capacity to revolutionize the tasks computers could perform. Because they establish new functionalities for users, radical innovations have strong potential to lead to significant growth in revenue and profits. Developing new processes is a critical part of producing radical innovations. Both types of innovation can create value, meaning that firms should determine when it is appropriate to emphasize either incremental or radical innovation. However, radical innovations have the potential to contribute more significantly to a firm's efforts to earn above-average returns.

Radical innovations are rare because of the difficulty and risk involved in developing them. The value of the technology and the market opportunities are highly uncertain. Because radical innovation creates new knowledge and uses only some or little of a firm's current product or technological knowledge, creativity is required. However, creativity does not create something from nothing. Rather, creativity discovers, combines, or synthesizes current knowledge, often from diverse areas. This knowledge is then used to develop new products that can be used in an entrepreneurial manner to move into new markets, capture new customers, and gain access to new resources. Such innovations are often developed in separate business units that start internal ventures.

Internally developed incremental and radical internal innovations result from deliberate efforts. These deliberate efforts are called internal corporate venturing, which is the set of activities firms use to develop internal inventions and especially innovations. Autonomous and induced strategic behaviors are the two types of internal corporate venturing. Each venturing type facilitates incremental and radical innovations. However, a larger number of radical innovations spring from autonomous strategic behavior while the greatest percentage of incremental innovations come from induced strategic behavior.

10. 6   Autonomous Strategic Behavior

Autonomous strategic behavior is a bottom-up process in which product champions pursue new ideas, often through a political process, by means of which they develop and coordinate the commercialization of a new good or service until it achieves success in the marketplace. A product champion is an organizational member with an entrepreneurial vision of a new good or service who seeks to create support for its commercialization. Product champions play critical roles in moving innovations forward. Indeed, in many corporations, "Champions are widely acknowledged as pivotal to innovation speed and success." The primary reason for this is that "no business idea takes root purely on its own merits; it has to be sold."Commonly, product champions use their social capital to develop informal networks within the firm. As progress is made, these networks become more formal as a means of pushing an innovation to the point of successful commercialization. Internal innovations springing from autonomous strategic behavior tend to diverge from the firm's current strategy, taking it into new markets and perhaps new ways of creating value for customers and other stakeholders.

Autonomous strategic behavior is based on a firm's wellsprings of knowledge and resources that are the sources of the firm's innovation. Thus, a firm's technological capabilities and competencies are the basis for new products and processes. GE depends on autonomous strategic behavior on a regular basis to produce innovations. Essentially, "the search for marketable services can start in any of GE's myriad businesses. An operating unit seeks out appropriate technology to better do what it already does. Having mastered the technology, it then incorporates it into a service it can sell to others." Changing the concept of corporate-level strategy through autonomous strategic behavior results when a product is championed within strategic and structural contexts. The strategic context is the process used to arrive at strategic decisions (often requiring political processes to gain acceptance). The best firms keep changing their strategic context and strategies because of the continuous changes in the current competitive landscape. Thus, some believe that the most competitively successful firms reinvent their industry or develop a completely new one across time as they compete with current and future rivals.

To be effective, an autonomous process for developing new products requires that new knowledge be continuously diffused throughout the firm. In particular, the diffusion of tacit knowledge is important for development of more effective new products. Interestingly, some of the processes important for the promotion of autonomous new product development behavior vary by the environment and country in which a firm operates. For example, the Japanese culture is high on uncertainty avoidance. As such, research has found that Japanese firms are more likely to engage in autonomous behaviors under conditions of low uncertainty.

10.7 Design for Six Sigma

It's the difference between getting a tune-up and a brand new engine; between patching your trousers and getting a new pair. Instead of constantly debugging products and processes that already exist - an effort that never ends, of course - DFSS starts from scratch to design the product or process to be virtually error free. This effectively replaces the usual trialand-error style with a cleaner, bump-free end result that also requires much less aftermarket tinkering. It's the classic 'pay me now or pay me later' solution, in which more time and effort are spent upfront so less will be spent after the fact.

Smart carpenters say, 'Measure twice and cut once.' And that's what Design for Six Sigma is about: getting it right the first time. If the design or process was flawed in the first place, you can only go so far with downstream fixes. In the case of producing products, manufacturing can only take quality away from the design, not improve it, so we must do our best to make the design as flawless as possible before we implement it.

Design for Six Sigma has already proven to be a groundbreaking strategic initiative for the corporations that have thoroughly implemented the methodology, and it's no exaggeration to say it has the potential to become the most significant management initiative of the 21st century. While such bold statements are often used to promote the latest hot idea, there are times, of course, when they are actually true. From the early returns of DFSS companies, it already appears that this could be one of those times.

Design for Six Sigma was created to enhance the one factor that almost every CEO has identified as the single sustainable competitive advantage: innovation. The problem is, it's also one of the most difficult things to manage. While all employees must adhere to certain guidelines to work for any corporation, creativity must be given the space to flourish. DFSS shows managers how to generate more creativity from their staff in a way that will not only preserve the integrity of the company but will actually strengthen it through better ideas, happier employees and environment that encourages growth instead of stifling it. Design for Six Sigma provides the means to accelerate innovation, which is why GE, Caterpillar, Delphi Automotive Systems, Dow Chemical and others have already entered the Design for Six Sigma race. Many others will follow, just as they have in pursuing Six Sigma. Those who excel in Design for Six Sigma will win; those who don't will face a perilous future.

Six is the sigma level of perfection we're aiming for. If your company's working at

One Sigma, for example, that means it's making about 700,000 defects per million opportunities (DPMO). At One Sigma you're getting things right only about 30 per cent of the time – a clearly unacceptable level of performance for everyone who doesn't play left field for the Yankees. Baseball is probably the only profession where a 30 per cent success rate is considered very good. Two Sigma is, obviously, better. If you're working at Two Sigma, you're making a little over 300,000 mistakes per million opportunities. In other words, you're batting about 70 per cent. Great for a major leaguer, but just OK in business. Most companies operate between Three and Four Sigma, which means they make between approximately 67,000 and 6,000 mistakes per million opportunities. If you're operating at 3.8 Sigma, that means you're getting it right 99 per cent of the time.

While this goal might seem impossible, there are actually companies out there that are consistently achieving between Five and Six Sigma quality. We'll discuss this more a little later, but the important point here is that they're not knocking themselves out to improve quality just for the sake of it. They're doing it to make more money by cutting costs and increasing profits. Most companies think improving quality costs money, so they see the quality-versus-profits balance as a trade-off, a tug-of-war between their customers and their accountants. They ask themselves, how much quality can we afford to give the customers and still make a profit? But Six Sigma companies flip that around. They've learned that quality saves money, because you have fewer throw-outs, fewer warranty payouts, fewer refunds and much higher rates of customer retention. And doing all that, in turn, increases profits.

Now that you've had a refresher course on the definition of Six Sigma, let's explore how it works. The power of Six Sigma is the combination of People Power with Process Power. The bulk of the work on People Power is done by middle management. A company's most outstanding people with proven drive and intellect are chosen to become Black Belts, a Six Sigma term denoting those who are most responsible for running Six Sigma projects. They are trained extensively in the Six Sigma philosophy and tools, and then given the support and resources they need to work full time on a specific project. Once the deadlines have been met and numerical goals have been reached, a Black Belt moves on to other projects. Process Power, on the other hand, encompasses five steps: define the problem, measure where you stand, analyze where the problem starts, improve the situation and control the new process to confirm that it's fixed. That boils down to a simple acronym, DMAIC, or as some people have learned to memorize it: dumb managers always ignore customers.

The idea behind Six Sigma is simple: instead of simply plugging leak after leak, the idea is to figure out why it's leaking and where, and attack the problem at its source. But Six Sigma does not address the original design of the product or process; it merely improves them. Design for Six Sigma is not simply a rehash of the lessons learned in Six Sigma but a fundamentally different methodology. DFSS complements the Six Sigma improvement methodology, but takes it one step further - or really, one step back - ferreting out the flaws of the product and the process during the design stage - not the quality control stage, or even the production stage. While Six Sigma focuses on improving existing designs, DFSS concentrates its efforts on creating new and better ones.

If your company were a house, it would work like this: while most business initiatives focus only on plugging leaky pipes and fixtures, a Six Sigma approach would examine the process and discover that the quality of the welding and sink taps was\ inadequate and replace them. DFSS would take one step further back in the process by designing the system - before it was ever installed - with welds and fixtures it knew would produce Six Sigma quality, without repairs or redesigns. Of course, few businesses involve leaky pipes. But all businesses involve customers - and understanding and pleasing them is the key, naturally, to business success. Traditionally, most companies have not taken the time or made the effort that is required to learn what their customers really want. DFSS requires applying resources to finding out what customers really want and then devoting the entire project to meeting the needs and desires of these customers. This works whether the customer is external - a car buyer, for example - or internal, such as the people in the accounting department. the production stage. While Six Sigma focuses on improving existing designs, DFSS concentrates its efforts on creating new and better ones.

10. 8　Well-structured problems and programmed decisions

Some problems are straightforward. The goal of the decision maker is clear, the problem is familiar, and information about the problem is easily defined and complete. Examples of these types of problems might include a customer wanting to return a purchase to a retail store, a supplier being late with an important delivery, a news team responding to an unexpected and fast-breaking event, or a university's handling of a student wanting to drop a class. Such situations are called structured problems since they are straightforward, familiar and easily defined. For instance, a server in a restaurant spills a drink on a customer's coat.

The manager has an upset customer and needs to do something. Because drinks are frequently spilled, there is probably some standardized routine for handling the problem. For example, the manager offers to have the coat cleaned at the restaurant's expense. This is what we call a programmed decision, a repetitive decision that can be handled by a routine approach. Because the problem is structured, the manager does not have to go to the trouble and expense of going through an involved decision process. Programmed decision making is relatively simple and tends to rely heavily on previous solutions.

The 'develop-the-alternatives' stage in the decision-making process either does not exist or is given little attention. Why? Because once the structured problem is defined, its solution is usually self-evident or at least reduced to very few alternatives that are familiar and that have proved successful in the past. The spilled drink on the customer's coat does not require the restaurant manager to identify and weight decision criteria or to develop a long list of possible solutions. Rather, the manager relies on a programmed decision, of which there are three types: a procedure, rule or policy.

A procedure is a series of sequential steps that a manager can use to respond to a structured problem. The only real difficulty is in identifying the problem. Once it is clear, so is the procedure. For instance, a purchasing manager receives a request from a sales manager for 15 Palm handhelds for the company's sales reps. The purchasing manager knows how to make this decision – follow the established purchasing procedure. A rule is an explicit statement that tells a manager what they can or cannot do. Rules are frequently used because they are simple to follow and ensure consistency. For example, rules about lateness and absenteeism permit supervisors to make disciplinary decisions rapidly and fairly.

Not all problems that managers face are well structured and solvable by a programmed decision. Many organizational situations involve unstructured problems, which are problems that are new or unusual and for which information is ambiguous or incomplete. For example, the selection of an architect to design a new corporate manufacturing facility in Bangkok is an example of a poorly structured problem. So too is the problem of whether to invest in a new, unproven technology or whether to shut down a money-losing division. When problems are unstructured, managers must rely on non-programmed decision making in order to develop unique solutions. Non-programmed decisions are unique and non-recurring and require custom-made solutions. When a manager confronts an unstructured problem, or one that is unique, there is no cut-and-dried solution. It requires a custom-made response through non-programmed decision making.

10.9   Risk analysis

A far more common situation is one of risk – those conditions in which the decision maker is able to estimate the likelihood of certain outcomes. The ability to assign probabilities to outcomes may be the result of personal experiences or secondary information. Under the conditions of risk, managers have historical data from personal experiences or secondary information that allow them to assign probabilities to different alternatives. Let us work through an example.

Suppose that you manage a ski resort in Queenstown, New Zealand. You are thinking about adding another lift to your current facility. Obviously, your decision will be significantly influenced by the amount of additional revenue that the new lift would generate, and additional revenue will depend on the level of snowfall. The decision is made somewhat clearer when you are reminded that you have reasonably reliable data on snowfall levels in your area. The weather data show that, during the past ten years, you had three years of heavy snowfall, five years of normal snow and two years of light snow. Can you use this information to determine the expected future annual revenue if the new lift is installed? If you have good information on the amount of revenue for each level of snow, the answer is 'yes'.

You can create an expected value formulation; that is, you can calculate the conditional return from each possible outcome by multiplying expected revenue by snowfall probability. The result is the average revenue you can expect over time if the given probabilities hold. As Table 7.6 shows, the expected revenue from adding a new ski lift is. Of course, whether that justifies a 'yes' or a 'no' decision depends on the costs involved in generating that revenue - factors such as the cost of building the lift, the additional annual operating expenses for another lift, the interest rate for borrowing money, and so forth.

What happens if you have a decision to make when you are not certain about the outcomes and cannot even make reasonable probability estimates? We call such a condition uncertainty. Managers do face decision-making situations of uncertainty. Under conditions of uncertainty, the choice between alternatives is influenced by the limited amount of information available to the decision maker.

When decision makers tend to think they know more than they do or hold unrealistically positive views of themselves and their performance, they are exhibiting the overconfidence bias. The immediate gratification bias describes decision makers who tend to want immediate rewards and to avoid immediate costs. For these individuals, decision choices that provide quick payoffs are more appealing than those in the future. The anchoring effect describes when decision makers fixate on initial information as a starting point and then, once set, fail to adjust adequately for subsequent information. First impressions, ideas, prices and estimates carry unwarranted weight relative to information received later. When decision makers selectively organize and interpret events based on their biased perceptions, they are using the selective perception bias. This influences the information they pay attention to, the problems they identify, and the alternatives they develop. Decision makers who seek out information that reaffirms their past choices and discount information that contradicts past judgments exhibit the confirmation bias.

These people tend to accept at face value information that confirms their preconceived views and are critical and sceptical of information that challenges these views. The framing bias is when decision makers select and highlight certain aspects of a situation while excluding others. By drawing attention to and highlighting specific aspects of a situation, while at the same time downplaying or omitting other aspects, they distort what they see and create incorrect reference points. The availability bias is when decision makers tend to remember events that are the most recent and vivid in their memory. The result? It distorts their ability to recall events in an objective manner and results in distorted judgments and probability estimates. When decision makers assess the likelihood of an event based on how closely it resembles other events or sets of events, that is the representation bias. Managers exhibiting this bias draw analogies and see identical situations where they do not exist.

The randomness bias is when decision makers try to create meaning out of random events. They do this because most decision makers have difficulty dealing with chance, even though random events happen to everyone and there is nothing that can be done to predict them. The sunk costs error is when decision makers forget that current choices cannot correct the past. They incorrectly fixate on past expenditures of time, money or effort in assessing choices, rather than on future consequences. Instead of ignoring sunk costs, they cannot forget them. Decision makers who are quick to take credit for their successes and to blame failure on outside factors are exhibiting the self-serving bias. Finally, the hindsight bias is the tendency for decision makers falsely to believe that they would have accurately predicted the outcome of an event once that outcome is actually known.

10. 10   Growth Models with Consumer Optimization

One finding will be that the saving rate is not constant in general but is instead a function of the per capita capital stock, k. Thus we modify the Solow – Swan model in two respects: first, we pin down the average level of the saving rate, and, second, we determine whether the saving rate rises or falls as the economy develops. We also learn how saving rates depend on interest rates and wealth and, in a later chapter, on tax rates and subsidies. The average level of the saving rate is especially important for the determination of the levels of variables in the steady state. In particular, the optimizing conditions in the Ramsey model preclude the kind of inefficient over saving that was possible in the Solow – Swan model.

The tendency for saving rates to rise or fall with economic development affects the transitional dynamics, for example, the speed of convergence to the steady state. If the saving rate rises with k, then the convergence speed is slower than that in the Solow –

Swan model, and vice versa. We find, however, that even if the saving rate is rising, the convergence property still holds under fairly general conditions in the Ramsey model. That is, an economy still tends to grow faster in per capita terms when it is further from its own steady-state position.

We show that the Solow - Swan model with a constant saving rate is a special case of the Ramsey model; moreover, this case corresponds to reasonable parameter values. Thus, it was worthwhile to begin with the Solow - Swan model as a tractable approximation to the optimizing framework. We also note, however, that the empirical evidence suggests that saving rates typically rise with per capita income during the transition to the steady state. The Ramsey model is consistent with this pattern, and the model allows us to assess the implications of this saving behavior for the transitional dynamics. Moreover, the optimizing framework will be essential in later chapters when we extend the Ramsey model in various respects and consider the possible roles for government policies. Such policies will, in general, affect the incentives to save.

The households provide labor services in exchange for wages, receive interest income on assets, purchase goods for consumption, and save by accumulating assets. The basic model assumes identical households—each has the same preference parameters, faces the same wage rate (because all workers are equally productive), begins with the same assets per person, and has the same rate of population growth. Given these assumptions, the analysis can use the usual representative-agent framework, in which the equilibrium derives from the choices of a single household. We discuss later how the results generalize when various dimensions of household heterogeneity are introduced.

Each household contains one or more adult, working members of the current generation. In making plans, these adults take account of the welfare and resources of their prospective descendants. We model this intergenerational interaction by imagining that the current generation maximizes utility and incorporates a budget constraint over an infinite horizon. That is, although individuals have finite lives, we consider an immortal extended family. This setting is appropriate if altruistic parents provide transfers to their children, who give in turn to their children, and so on. The immortal family corresponds to finite-lived individuals who are connected through a pattern of operative intergenerational transfers based on altruism.

We now assess the convergence properties of the model with a second approach, which uses numerical methods to solve the nonlinear system of differential equations. This approach avoids the approximation errors inherent in linearization of the model and provides accurate results for a given specification of the underlying parameters. The disadvantage is the absence of a closed-form solution. We have to generate a new set of answers for each specification of parameter values.

Our analysis thus far has considered a single household as representing the entire economy. The consumption and saving decisions of the representative agent are supposed to capture the behavior of the average agent in a complex economy with many families. The important question is whether the behavior of this "representative" or "average" household is really equivalent to what we would get if we averaged the behavior of many heterogeneous families.

Although the transitional dynamics is complicated, it is straightforward to work out the characteristics of the steady state. The key point is that, in a steady state, an increase in household assets would be used to raise consumption uniformly in future periods. This property makes it easy to compute propensities to consume for future periods with respect to current assets and, therefore, makes it easy to find the first-order optimization condition for current consumption. Only the results are presented here.

参 考 文 献

[1] Frederic Frery. The Fundamental Dimensions of Strategy [J]. MIT Sloan Management Review，2006(Fall)，71—75.

[2] Ghemawat Pankaj. Competition And Business Strategy In Historical Perspective[J]. Business History Review,2002(76):37—74.

[3] Hambrick D. C. , Fredrickson J. W.. Are You Sure You Have a Strategy? [J]. Academy of management executive,2001,15(4):48—59.

[4] Hoskisson R. E. , Hitt M. A. And Wan W. P.. Theory And Research In Strategic Management :Swings Of A Pendulum[J]. Journal Of Management,1999, 25(3):417—456.

[5] Mir Raza,Watson Andrew. Strategic management and the philosophy of science: the case for a constructivist methodology [J]. Strategic Management Journal 2000(21): 941—953.

[6] Michael A. Hitt, R. Duane Ireland, Robert E. Strategy Management: Competitiveness and Globalization[M]. China Machine Press, 2002.

[7] Peteraf,Margaret A. The Cornerstones of Competitive Advantage: A Resource—Based View[J]. Strategic Management Journal 1993(14):179—91.

[8] Porter, Michael E. Towards a Dynamic Theory of Strategy, Strategic Management [J]. Journal 1991. (Summer Special Issue) 12: 95—117.

[9] Poter, M. E. What's strategy[J]. Harvard Business Review,1996(6):61—78.

[10] Raphaei Amit, Christoph Zott. Value creation in e—business [J]. Strategic Management Journal, 2001(22):493—520.

[11] Rumelt R. P. ,Schendel D. And Teec D. J. Strategic Management And Economics [J]. Strategic Management Journal,1991(12):5—29.

[12] Shobha S. Das, Andrew H. Van de Ven . Competing with new product technologies: a process model of strategy[J]. Management Science,2000,46(10): 1300—1316.

[13] Stephen P Robbins. Management (Fourth Edition) [M]. Prentice Hall Press,1998.

[14] Yadong luo, Seung No Park. Strategic alignment and performance of market—seeking MNCs in china[J]. Strategic Management Journal,2001(22):141—155.

[15] Yiannis E. Spanos, Spyros Lioukas. An examination into the causal logic of rent generation: contrasting Porter's competitive strategy framework and the resource—

based perspective[J]. Strategic Management Journal，2001(22)：907—934.

［16］戴伊，雷布斯坦因.动态竞争战略[M].上海：上海交大出版社,2003.

［17］亨利·明茨伯格,魏江译.战略历程[M].机械工业出版社,2007.

［18］坎贝尔,卢斯.核心能力战略[M].大连：东北财经大学出版社,2003.

［19］康健,胡祖光.产业集群脆弱性测度模式初探[J].经济地理,2012,32(2)：111—115.

［20］康健,胡祖光.产业集群情境下出版企业供应链分析[J].中国出版,2012,(8)：17—20.

［21］康健.服务企业虚拟经营战略实施途径研究——基于价值链的视角[J].河南社会科学,2011,(2)：163—165.

［22］康健,胡祖光.基于区域产业互动的三螺旋协同创新能力评价研究[J].科研管理,2014,35(5)：19—26.

［23］康健.基于物元分析的虚拟服务企业风险评估模型研究[J].科技管理研究,2009,(4)：183—185.

［24］康健.价值链重构视角下服务企业虚拟经营战略研究[J].中国商贸,2010,(8)：14—15.

［25］康健.竞争情报系统在战略资源评价中的应用研究[J].情报杂志,2007,(11)：20—21,25.

［26］康健.论服务企业虚拟经营战略的研究视角[J].企业经济,2009,(5)：70—72.

［27］康健.培育我国出版产业集群的竞争力研究[J].编辑之友,2013,(10)：33—36.

［28］康健.新经济环境下企业战略管理中的关键点控制研究[J].经济视角,2010,(5)：20—21.

［29］康健.战略性新兴产业集群化发展的供应链支持网络绩效提升研究———以衡阳市文化创意产业为例[J].经济视角,2012,(12)：19—22.

［30］康健,胡祖光.战略性新兴产业与生产性服务业协同创新研究：演化博弈推演及协同度测度[J].科技管理研究,2015,35(4)：154—161.

［31］理查德·惠廷顿,王智慧译.战略是什么[M].中国劳动社会保障出版社,2004.

［32］李兴旺.竞争层次论[M].内蒙古大学出版社,2005.

［33］李兴旺,何辛锐.战略管理理论范式研究[J].管理学家(学术版),2009(4)：12—18.

［34］李烨.从企业管理主题变迁透析战略管理理论的新发展[J].华东经济管理,2008(10)：113—116.

［35］李玉刚.战略管理研究[M].华东理工大学出版社,2005.

［36］林健,李焕荣.企业战略管理理论核心逻辑分析[J].经济管理,2002(22)：18—23.

［37］刘军,吴维库.战略的矛盾逻辑[J].企业管理,2004(9)：88—89.

［38］洛斯·加里洛.战略逻辑[M].经济管理出版社,2005.

［39］梅森·卡彭特,杰瑞德·桑德斯,王迎军译.战略管理：动态观点[M].机械工业出版社,2009.

［40］彭勇涛.企业战略逻辑分析[J].科技管理研究,2002(6)：58—61.68.

[41] 盛昭瀚,蒋德麟.演化经济学[M].上海:上海三联书店,2002.

[42] 孙中伟,宋仁慧.战略思考的系统与动态逻辑探析[J].生产力研究,2008(1):75—78.

[43] 石军伟.顾客价值、战略逻辑创新与企业核心竞争力[J].贵州财经学院学报,2002 (3):5—9.

[44] 石军伟.企业战略理论的逻辑比较[J].经济管理,2002(8):47—53.

[45] 石盛林.战略管理理论演变:基于企业理论视角的回顾[J].科技进步与对策,2010(4): 156—159.

[46] 唐欣.基于 BP 神经网络模型的企业绿色经营绩效评价方法[J].统计与决策,2012, (2):87—88.

[47] 唐欣.循环经济视角下企业绩效三重评价模型的构建[J].企业经济,2009,(6):38 —41.

[48] 唐欣.构建企业环境会计报告模式的设想[J].商业会计,2008,(16):16—17.

[49] 唐欣.基于模糊数学评价的环境投资项目绩效审计模式研究[J].财会研究,2010, (4):47—49.

[50] 唐欣.竞争情报系统在企业环境价值链重构中的应用研究[J].财政监督,2009,(12): 14—15.

[51] 唐欣.作业成本法在环境成本核算中的应用——基于啤酒生产企业的分析[J].财会 通讯,2009,(2):54—55.

[52] 唐欣.基于环境价值链的环境成本计量模型探讨[J].财会通讯,2008,(12):48—49.

[53] 唐欣.对真实性原则的几点思考[J].山西财经大学学报,2005,(2).

[54] 唐欣.AHP 方法在战略管理会计竞争价值链分析中的应用[J].财会通讯,2006,(4).

[55] 汪红丽.论战略价值化管理及其价值创新战略逻辑[J].价值工程,2005(8):19—22.

[56] 王永贵.21 世纪企业制胜方略—构筑动态竞争优势[M].北京:机械工业出版社,2001.

[57] 项国鹏.企业战略管理范式的转型[J].经济管理.新管理,2001(16):4—11.

[58] 杨林,陈传明.国外企业战略管理理论演变:矛盾论的视角[J].经济管理,2005(1):14 —18.

[59] 杨逸、贾良定、陈永霞.认知学派:战略管理理论发展前沿[J].南大商学评论,2007,4 (15):178—194.

[60] 伊藤诚,阿尔布里坦,左格等.资本主义的发展阶段[M].北京:经济科学出版社,2003.

[61] 殷瑾,陈劲.顾客价值创新的战略逻辑和基本模式[J].科研管理,2002(5):110—114.

[62] 张平.科学发展理性:中国转型社会的公共理性[J].湖南科技大学学报(社会科学版) ,2012,(4):54—57.

[63] 张平.中国转型社会公共理性特征论[J].湘潭大学学报(哲学社会科学版),2012, (4):20—23.

[64] 张平.衡阳市产业园区发展的现状及对策分析[J].经济研究导刊,2012,(2):81— 82,124.

[65] 张平.衡阳市"十二五"现代产业体系的构建研究[J].山东省农业管理干部学院学

报,,2012,(1):51—53.

[66] 张平. 湖南中小企业集群发展研究[J].经济地理,2006,(6):1001—1104.

[67] 张平. 关于衡阳民营企业的发展[J]. 天津市经理学院学报,2006,(3):21—22.

[68] 张平. 中国中部地区中小企业集群发展研究[J]. 生产力研究,2007,(8):128—133.

[69] 芝加哥大学商学院等.把握战略—MBA 战略精要[M].北京:北京大学出版社,2003.

[70] 周三多. 战略管理思想史 [M].上海复旦大学出版社,2002.

后　　记

　　新经济时代中,现代企业的发展和生存环境都发生了巨大的变化,许多管理理念和方法在不断地被实践所扬弃。为了更为准确地把握新经济发展的脉搏,企业管理者有必要针对动态变化的环境对自身进行准确的战略规划、战略制订、战略实施和战略控制。因此,战略管理即成为工商管理专业的核心教学模块之一。

　　在本书中,编者们已经竭尽所能将战略管理学科的方法、工具和思路展示给读者。但是当这本书稿完成时,依然深感不安,因为离我们的目标还有很大差距,尤其是新经济环境下战略管理运行和实施的新理念和方法在企业经营活动中灵活应用的模式问题。

　　本书的完成得到了许多方面的帮助。首先要感谢湖南工学院分管教学工作的张平校长,经济管理学院杨凤鸣院长、赵少平副院长,是他们为本书的撰写创造了非常好的学习和工作条件;其次要感谢经济管理学院的学术先行者陈国生教授;最后还要感谢管理教研室的同事袁鹏老师、黄飞老师、刘秋英老师、祁德军老师、蒋晓林老师、肖晓峰老师、张桂华老师、王长富老师、唐跃文老师、范文峰老师、陈杰老师、陈晓亮老师、罗白璐老师、唐婧老师、陆利军老师、廖珊老师、刘锦志老师等。

　　谨以本书作为我们近几年课堂教学和教研教改工作的一个阶段性总结,同时也作为以后继续研究的一个起点。

本书所有编者